Resolving Conflicts Arising from the Privatization of Environmental Data

Committee on Geophysical and Environmental Data

Board on Earth Sciences and Resources

Division on Earth and Life Studies

National Research Council

NATIONAL ACADEMY PRESS
Washington, D.C.

NOTICE: The project that is the subject of this report was approved by the Governing Board of the National Research Council, whose members are drawn from the councils of the National Academy of Sciences, the National Academy of Engineering, and the Institute of Medicine. The members of the committee responsible for the report were chosen for their special competences and with regard for appropriate balance.

This study was supported by the federal agencies of the U.S. Global Change Research Program (USGCRP) through the National Oceanic and Atmospheric Administration (NOAA) under Contract No. 50-DKNA-7-90052. The opinions, findings, conclusions, and recommendations expressed herein are those of the authors and do not necessarily reflect the view of NOAA, USGCRP, or any of its sub-agencies.

International Standard Book Number 0-309-07583-1

Additional copies of this report are available from:

National Academy Press
2101 Constitution Avenue, N.W.
Box 285
Washington, DC 20055
800-624-6242
202-334-3313 (in the Washington metropolitan area)
http://www.nap.edu

Cover: Banyan tree representing the parts of environmental information systems: an extensive root system that draws data from many sources; a trunk in which information is synthesized into core information products; multiple branches that distribute and enhance the core products into value-added products; and leaves, which represent different uses of information products of the tree. As with Banyan trees, roots, trunk, and branches are interconnected with one another and with other information system trees. Illustration courtesy of Van Nguyen, National Academy Press.

Copyright 2001 by the National Academy of Sciences. All rights reserved.

Printed in the United States of America

THE NATIONAL ACADEMIES

National Academy of Sciences
National Academy of Engineering
Institute of Medicine
National Research Council

The **National Academy of Sciences** is a private, nonprofit, self-perpetuating society of distinguished scholars engaged in scientific and engineering research, dedicated to the furtherance of science and technology and to their use for the general welfare. Upon the authority of the charter granted to it by the Congress in 1863, the Academy has a mandate that requires it to advise the federal government on scientific and technical matters. Dr. Bruce M. Alberts is president of the National Academy of Sciences.

The **National Academy of Engineering** was established in 1964, under the charter of the National Academy of Sciences, as a parallel organization of outstanding engineers. It is autonomous in its administration and in the selection of its members, sharing with the National Academy of Sciences the responsibility for advising the federal government. The National Academy of Engineering also sponsors engineering programs aimed at meeting national needs, encourages education and research, and recognizes the superior achievements of engineers. Dr. Wm. A. Wulf is president of the National Academy of Engineering.

The **Institute of Medicine** was established in 1970 by the National Academy of Sciences to secure the services of eminent members of appropriate professions in the examination of policy matters pertaining to the health of the public. The Institute acts under the responsibility given to the National Academy of Sciences by its congressional charter to be an adviser to the federal government and, upon its own initiative, to identify issues of medical care, research, and education. Dr. Kenneth I. Shine is president of the Institute of Medicine.

The **National Research Council** was organized by the National Academy of Sciences in 1916 to associate the broad community of science and technology with the Academy's purposes of furthering knowledge and advising the federal government. Functioning in accordance with general policies determined by the Academy, the Council has become the principal operating agency of both the National Academy of Sciences and the National Academy of Engineering in providing services to the government, the public, and the scientific and engineering communities. The Council is administered jointly by both Academies and the Institute of Medicine. Dr. Bruce M. Alberts and Dr. Wm. A. Wulf are chairman and vice chairman, respectively, of the National Research Council.

COMMITTEE ON GEOPHYSICAL AND ENVIRONMENTAL DATA

J. BERNARD MINSTER, *Chair*, Scripps Institution of Oceanography, University of California, La Jolla
FRANCIS P. BRETHERTON, University of Wisconsin, Madison (*Chair* through 2000)
DAVID H. BROMWICH, Ohio State University, Columbus
MARY ANNE CARROLL, University of Michigan, Ann Arbor
JEFF DOZIER, University of California, Santa Barbara
DAVID M. GLOVER, Woods Hole Oceanographic Institution, Woods Hole, Massachusetts
GEORGE H. LEAVESLEY, U.S. Geological Survey, Denver, Colorado
MARK J. McCABE, Georgia Institute of Technology, Atlanta
JOHN M. MELACK, University of California, Santa Barbara
JOYCE E. PENNER, University of Michigan, Ann Arbor (through 2000)
ROY RADNER, New York University, New York
CYNTHIA E. ROSENZWEIG, NASA Goddard Institute for Space Studies, New York, New York (through 1999)
WILLIAM F. RUDDIMAN, University of Virginia, Charlottesville
ROBERT J. SERAFIN, National Center for Atmospheric Research, Boulder, Colorado
CARL WUNSCH, Massachusetts Institute of Technology, Cambridge (through 2000)

National Research Council Staff

ANNE M. LINN, Senior Program Officer
JENNIFER T. ESTEP, Administrative Associate
SHANNON L. RUDDY, Project Assistant

BOARD ON EARTH SCIENCES AND RESOURCES

RAYMOND JEANLOZ, *Chair*, University of California, Berkeley
JOHN J. AMORUSO, Amoruso Petroleum Company, Houston, Texas
PAUL B. BARTON, JR., U.S. Geological Survey (emeritus), Reston, Virginia
DAVID L. DILCHER, University of Florida, Gainesville
BARBARA L. DUTROW, Louisiana State University, Baton Rouge
ADAM M. DZIEWONSKI, Harvard University, Cambridge, Massachusetts
WILLIAM L. GRAF, Arizona State University, Tempe
GEORGE M. HORNBERGER, University of Virginia, Charlottesville
SUSAN W. KIEFFER, S.W. Kieffer Science Consulting, Inc., Bolton, Ontario, Canada
DIANNE R. NIELSON, Utah Department of Environmental Quality, Salt Lake City
JONATHAN G. PRICE, Nevada Bureau of Mines & Geology, Reno
BILLIE L. TURNER II, Clark University, Worcester, Massachusetts

National Research Council Staff

ANTHONY R. DE SOUZA, Director
TAMARA L. DICKINSON, Senior Program Officer
DAVID A. FEARY, Senior Program Officer
ANNE M. LINN, Senior Program Officer
PAUL M. CUTLER, Program Officer
LISA M. VANDEMARK, Program Officer
KRISTEN L. KRAPF, Research Associate
KERI H. MOORE, Research Associate
MONICA R. LIPSCOMB, Research Assistant
JENNIFER T. ESTEP, Administrative Associate
VERNA J. BOWEN, Administrative Assistant
YVONNE P. FORSBERGH, Senior Project Assistant
KAREN L. IMHOF, Senior Project Assistant
SHANNON L. RUDDY, Project Assistant
TERESIA K. WILMORE, Project Assistant
WINFIELD SWANSON, Editor

Acknowledgments

This report has been reviewed in draft form by individuals chosen for their diverse perspectives and technical expertise, in accordance with procedures approved by the NRC's Report Review Committee. The purpose of this independent review is to provide candid and critical comments that will assist the institution in making its published report as sound as possible and to ensure that the report meets institutional standards for objectivity, evidence, and responsiveness to the study charge. The review comments and draft manuscript remain confidential to protect the integrity of the deliberative process. We wish to thank the following individuals for their review of this report:

Steven T. Berry, Yale University
Robert F. Brammer, The Analytical Sciences Corporation
Otis B. Brown Jr., University of Miami
Inez Y. Fung, University of California, Berkeley
Thomas M. Holm, U.S. Geological Survey, EROS Data Center
John A. Knauss, University of Rhode Island, and Scripps Institution of Oceanography
Harlan J. Onsrud, University of Maine
Carol A. Wessman, University of Colorado, Boulder

Although the reviewers listed above have provided many constructive comments and suggestions, they were not asked to endorse the conclusions or recommendations nor did they see the final draft of the report before its release. The review of this report was overseen by Freeman Gilbert and William L. Chameides, appointed by the National Research Council, who were responsible for making certain that an independent examination of this report was carried out in accordance with institutional procedures and that all review comments were carefully considered. Responsibility for the final content of this report rests entirely with the authoring committee and the institution.

Preface

The Earth's atmosphere, oceans, and biosphere form an integrated system that transcends national boundaries. To understand the elements of the system, the way they interact, and how they have changed with time, it is necessary to collect and analyze environmental data from all parts of the world. Consequently, over the past 100 years international programs for global change research and environmental monitoring have relied on policies guaranteeing full and open access to data (i.e., data and information are made available without restriction, on a non-discriminatory basis, for no more than the cost of reproduction). Such policies have facilitated significant scientific discoveries, as well as informed public policy.

However, the commercialization of government agencies in Europe and Canada, coupled with the rise of the commercial remote sensing industry, is changing the nature of the environmental science enterprise. Commercialized government agencies and private-sector organizations typically sell and/or restrict environmental data, making it difficult to collect and exchange the information upon which society depends. This report, which was requested by the U.S. Global Change Research Program,[1] identifies the issues and conflicts that arise from the different goals and objectives of the stakeholders in environmental information—

---

[1] The U.S. Global Change Research Program is a multi-agency program aimed at "providing a sound scientific understanding of the human and natural forces that influence the Earth's climate system—and thus provide a sound scientific basis for national and international decision making on global change issues." Nine federal agencies formally participate in the program. See Subcommittee on Global Change Research, 2000, *Our Changing Planet: The FY 2001 U.S. Global Change Research Program*, White House Office of Science and Technology Policy, Washington, D.C., 74 pp.

scientists, private-sector organizations, government agencies, policy makers, and the general public. The charge to the committee was to

> examine the impact of commercialization policies (including database legislation) on established scientific practices, including data publication, use of data for multiple purposes, data sharing, and noncommercial research in the private sector. Particular emphasis will be placed on policies concerning global or regional environmental science, including atmospheric, oceanic, solid-earth, biospheric, and polar science. Examples of data restrictions encountered by scientists and scientific data centers will be used to illustrate (1) problems in obtaining, using, sharing, or publishing data and (2) solutions that have worked in the past.

The Committee on Geophysical and Environmental Data has been concerned with national and international data policy for 34 years. This report builds on previous reports that examine the impact of proposed and existing data policies on the environmental science community.[2]

In gathering information for this report, the committee solicited input from intellectual property lawyers, private-sector organizations that collect environmental data or create value-added data products, economists, federal government agencies and data centers, international science organizations, and the broader scientific community. Altogether, six meetings were held to gather information and analyze the results.

Information was also gathered from the literature and World Wide Web sites. The information from Web sites provided in this report was correct, to the best of our knowledge, at the time of publication. It is important to remember, however, the rapidly changing content of the Internet. Resources that are free and publicly available one day may require a fee or restrict access the next, and the location of items may change as menus and home pages are reorganized.

The committee acknowledges the following individuals who briefed the committee or provided other input: Prue Adler, Jon Band, Roger Barry, Bruce Bauer, Mary Case, Robert Chen, Donald Collins, John Curlander, William Draeger, Bolling Farmer, Wanda Ferrell, Michael Freilich, Lee Fu, Steven Goodman, Richard Greenfield, Allen Hittelman,

---

[2]For example, see NRC, 1995, *On the Full and Open Exchange of Scientific Data*. National Research Council, Washington, D.C., 21 pp.

Thomas Holm, Peter Jaszi, Bruce Joseph, Thomas Kalvelage, Thomas Karl, Verne Kaupp, Steven Kempler, Joseph King, Herbert Kroehl, Michael Loughridge, Steve Maurer, Richard McGinnis, B. Greg Mitchell, Stephen Nelson, Kurt Schnebele, Suzanne Scotchmer, Mark Seeley, George Sharman, John Shaw, August Shumbera, Kurtis Thome, Shelby Tilford, Ferris Webster, Peter Weiss, Stanley Wilson, Robert Winokur, David Wolf, and James Yoder.

Finally, the committee is deeply indebted to the staff of the National Research Council, notably Jenny Estep and Shannon Ruddy, for arranging the numerous meetings required to obtain the viewpoints of the communities involved in this study. We especially thank the study director, Anne Linn, for her steadfast support of committee activities, for her willpower to keep us "on target" and, ultimately, for taking charge of the main report production effort, leading the interminable succession of drafts to successful convergence.

Francis Bretherton
Past Chair

Bernard Minster
Current Chair

Contents

EXECUTIVE SUMMARY 1

1 INTRODUCTION 7
   A Changing World, 7
   The Information Tree, 11
   Organization of the Report, 13

2 STAKEHOLDER VIEWPOINTS 15
   Scientist Views, 15
   Private-Sector Views, 19
   Government Agency Views, 23
   Policy Maker Views, 25
   General Public Views, 27

3 ENVIRONMENTAL INFORMATION SYSTEMS 29
   The Environmental Information System Tree, 31

4 POLICY AND ECONOMIC FRAMEWORK FOR PUBLIC-
   PURPOSE ENVIRONMENTAL INFORMATION SYSTEMS 37
   Data Policy, 37
   Compatibility of Open Access With a Competitive Market, 40
   Requirements of Public-Purpose Environmental Information
      Trees, 49

5 WAR AND PEACE AMONG STAKEHOLDERS 53
   Information Systems Created Purely for Public Purposes, 54
   Information Systems and Public-Private Partnerships, 68
   Overall Lessons Learned, 72
   The Need for a Process of Negotiating Among Stakeholders, 73

6 RECONCILING THE VIEWS OF THE STAKEHOLDERS 75
   Guidelines for Interactions Between Scientists and Private-
      Sector Organizations, 75
   Guidelines for Interactions Between Government Agencies
      and Private-Sector Organizations, 79

APPENDIXES 89

A Scientific Practices 91

B Intellectual Property Rights to Data 95

C Acronyms 99

Executive Summary

THE NEED FOR RELIABLE INFORMATION

Reliable collections of science-based environmental information are vital for many groups of users and for a number of purposes. For example, electric utility companies predict demand during heat waves, structural engineers design buildings to withstand hurricanes and earthquakes, water managers monitor each winter's snow pack, and farmers plant and harvest crops based on daily weather predictions. Understanding the impact of human activities on climate, water, ecosystems, and species diversity, and assessing how natural systems may respond in the future are becoming increasingly important for public policy decisions. Environmental information systems gather factual information, transform it into information products, and distribute the products to users. Typical uses of the information require long-term consistency; hence the operation of the information system requires a long-term commitment from an institution, agency, or corporation. The need to keep costs down provides a strong motivation for creating multipurpose information systems that satisfy scientific, commercial and operational requirements, rather than systems that address narrow objectives. This report focuses on such shared systems.

The five stakeholder groups in shared environmental information systems—research scientists, private-sector organizations, government agencies, policy makers, and the general public—have different goals and modes of operation. In particular, public-sector users (scientists, government agencies, and policy makers) generally rely on full and open access to data (i.e., data are made available without restriction for any use for no more than the marginal cost of filling a user request). On the other hand, in order to generate a financial return most private-sector organizations (for-profit producers and distributors of data and products) must restrict access to data. If the price of data increases without a commensurate increase in scientific value, then scientific and technical progress will decline. Nevertheless, some private-sector entities and Congress are urging government agencies to increase the involvement of the private sector in collecting and disseminating data and creating data products for public purposes. A number of government-developed satellite technologies are sufficiently mature to permit private-sector companies to enter the remote-sensing industry. Once established, however, private-sector organizations often do not want the government to compete with them by continuing to collect observations or to produce and disseminate information products. This view has been echoed by Congress in bills authorizing funding for federal agencies and in legislation forbidding agencies from competing with the private sector. These and other stakeholder viewpoints must be reconciled for the system to work. Underlying such a reconciliation should be the principle that the public welfare is best served by information systems that establish the relevant facts and enable the widest distribution to the public of facts and knowledge derived from them. Establishing the facts and distributing information are distinct functions that warrant separate consideration.

Recommendation. Environmental information systems that are created by the U.S. government to serve a public purpose should continue to establish facts that are accessible to all. To facilitate further distribution these facts should be made available at no more than the marginal cost of reproduction and should be useable without restriction for all purposes.

This recommendation extends the current practice of supplying most environmental data free or at marginal cost. U.S. policy (OMB Circular A-130) specifies that data should be made available at no more than incremental cost, which is slightly higher.

Given the above recommendation, the question is what roles can the private sector play effectively in shared-use, public-purpose environmental information systems?

THE ENVIRONMENTAL INFORMATION SYSTEM TREE

Environmental information systems created for public purposes can be portrayed in terms of a simple analogy—an information tree consisting of four parts:

1. An extensive root system that draws data from many different sources and organizations.
2. A trunk in which all available information is synthesized into a limited set of core products.
3. Multiple branches that distribute and enhance the core products into value-added products, each branch serving a distinct community of users.
4. Leaves, which represent uses of information products of the trunk and branches.

Collecting measurements and developing core products are typically the most expensive parts of the information system. It is not possible for private-sector organizations to recoup these costs solely by selling information products at the marginal cost of reproduction. Because of the cost structure of public-purpose environmental information systems and the need for their scientific validity, the public interest is best served by funding the trunk and roots out of taxpayer resources, and providing full and open access to a set of reliable core products. However, the value of distributing information derived from the core products in a convenient form to as broad a group of users as possible (i.e., through the branches) has made some privately-driven value-added markets successful.

Recommendation. The practice of public funding for data collection and synthesis should continue, thereby focusing contributions of the private sector primarily on value-added distribution and specific observational systems.

If private-sector organizations are able to provide a stable supply of high-quality data that fulfills public purposes without compromising the commercial market, then data collection in public-purpose information systems can in principle be privatized or managed through public-private partnerships. Similarly, the marketplace may provide an appropriate mechanism for deciding what value-added products are developed, although the government may have to provide them if commercial value-added products are not suitable for public purposes. The decision on public versus private funding should not be an ideological one. Rather, the choice of whether to acquire data or value-added products to meet government missions and mandates by direct funding or to purchase them through private-sector initiative must be based on sound analysis of the value of the information to the public good, likely market forces, revenues, and costs. The government should not expect the commercial market to supply data or value-added products on a full and open basis. Thus, commercial data or information products meant specifically to meet public-sector needs should be purchased and wholly owned by the government and placed in the public domain.

A PROCESS FOR NEGOTIATING AMONG STAKEHOLDERS

The objectives of the information system broadly constrain the roles of the five stakeholder groups. However, there is currently no recognized process for the stakeholders or their representatives to negotiate solutions that optimize common interests and minimize conflicts. Such a process is particularly important for information systems created with a mixture of public and private objectives because virtually every aspect of the system is negotiable. Issues must be resolved at the policy level (e.g., public funding of the trunk) and in the implementation details (e.g., the priorities for core products). Solutions will depend on the particular circumstances of the information system at hand. Thus, policy makers cannot expect to write a general rule that will settle conflicts for all stakeholders in all situations. Yet, finding common ground is imperative if the nation is to benefit from using environmental resources in a sustainable fashion and humankind is to face the challenges associated with their impact on the environment.

Recommendation. U.S. federal agencies with responsibility for multi-purpose environmental information systems should establish a clear, visible process by which representatives of all the stakeholder groups discuss the performance and negotiate the redesign of such systems with the goal of reconciling their interests.

COMMERCIALIZATION AND PRIVATIZATION

The terms "commercialization" and "privatization" are commonly used interchangeably, yet they mean different things and have different implications for public-purpose information systems. Commercialization is defined in this report to mean the financial exploitation of government data, whereas privatization refers to the transfer of government functions to the private sector. In the United States, public laws providing for unrestricted, affordable access to government data permit the coexistence of commercial exploitation of government data with public-sector uses, such as scientific research. As a result, commercialization maximizes the use and thus the value of data to all users. The same is not true in countries that exercise intellectual property rights over government data and thus limit the extent to which government-collected data can be used, even in international collaborations. By making it more difficult to integrate global datasets and share knowledge, such a commercialization policy will fail to achieve the maximum benefits provided by international collaboration and the scientific endeavor.

Privatization is not without risk to the public because it involves discontinuing government functions with proven value in favor of private-sector services for which benefits may never accrue. The risks are greatest in data collection because of the potential for price increases that disrupt scientific practices or gaps in the long-term record of environmental change. Nevertheless, under certain conditions, the collection of data and/or generation of data products can be transferred beneficially from the government to the private sector. In fact, by failing to do so, the full public benefit may not be achieved.

Decisions concerning which functions should be public and which ones should be private must be made case by case. Most decisions will involve the transfer of government functions to the private sector, but some will concern re-entry of the government as a supplier.

Recommendation. Before transferring government data collection and product development to private-sector organizations, the U.S. government should ensure that the following conditions will be satisfied: (1) avoidance of market conditions that give any firms significant monopoly power; (2) preservation of full and open access to core data products; (3) assurance that a supply of high-quality infor-

mation will continue to exist; and (4) minimized disruption to ongoing uses and applications.

1

Introduction

A CHANGING WORLD

Whether it is the air we breathe, the flowers we look forward to each spring, or the tornado that threatens our houses, the natural environment concerns us all. In our industrialized society, we count increasingly on reliable, factual information (see Box 1.1 for definitions) about the environment. Electric utility companies predict demand during heat waves; structural engineers design buildings to withstand hurricanes and earthquakes; water managers monitor each winter's snow pack. Over the past several decades it has become increasingly apparent that humans are altering climate all over the globe, whole ecosystems are being transformed, and innumerable species are becoming extinct. The implications of such fundamental changes are largely unknown, bringing a new urgency to understanding how natural systems have fared under external stresses in the past, documenting how they are responding at present, and establishing the scientific principles that allow us to predict their possible future courses.

All these applications—historical, current, or predictive—depend on science-based measurements, gathered and analyzed within a formal or informal information system framework. In many cases, an established process exists for exchanging, compiling, and interpreting environmental data nationally and internationally. In other cases, the process is less structured, but it still takes place through research and publication in scientific journals and government statistics or reports. These data are then correlated and interpreted by scientists and engineers to provide a reliable, factual basis for actions by others.

To further environmental understanding and develop good public policies for dealing with all aspects of the environment, the U.S.

government invests in basic research[1] and in information systems. However, declining agency budgets, which force the government to seek partners for sharing costs, as well as improvements in technology for collecting, handling, processing, and publishing data have opened new opportunities for the private sector to participate in the environmental enterprise. Although the involvement of the private sector brings potential advantages to science and society, it also introduces laws and business practices that are different from those of the scientific community. For example, international programs for global change research and environmental monitoring depend on policies guaranteeing full and open access to data (i.e., data and information made available without restriction on a nondiscriminatory basis for no more than the cost of reproduction and distribution). However, the private sector and commercialized government agencies in other countries operate in a commercial environment in which revenues must at least cover the costs of generating a data product, and controlling access to data is key to remaining competitive.

Five major groups of stakeholders, each of which has different goals, generate and/or use environmental information:

1. scientists involved in generating and interpreting data;
2. government agencies involved in funding much of the enterprise and in delivering products that achieve the overall goals of understanding the environment and providing information to improve decision making concerning the environment;
3. private-sector organizations, which have an increasing role in collecting data and producing value-added products;
4. policy makers, who make informed judgments about what is in the best long-term interests of the communities they represent; and
5. the general public, in whose interest basic research and environmental monitoring are being undertaken.

The purpose of this report is to identify the issues and potential conflicts that inevitably arise from interactions among these five groups. Special attention is given to the concerns of scientists, who currently

---

[1] Governments also invest in basic research because it yields enormous economic benefits. See C.I. Jones, and J.C. Williams, 1998, Measuring the Social Return to R&D, *Quarterly Journal of Economics*, v. 113(4), p. 1119-1135.

enjoy full and open access to an enormous quantity of government-collected data. A shift from public funding to a commercial market could relegate scientists' uses to a small niche,[2] giving scientists little or no voice in the collection of data that are necessary for understanding the environment and for generating knowledge on behalf of the public. Unless accomplished under carefully crafted conditions, such a shift with its associated increase in prices (compared with marginal cost) and restrictions on use could disrupt or even fundamentally change the scientific practices that have led to the scientific and economic successes of the last half century.

The Committee on Geophysical and Environmental Data was charged with examining the impact of commercialization and privatization policies (including database legislation) on established scientific practices in the environmental sciences (ocean, atmosphere, land surface, solid-earth), with an emphasis on (1) problems in obtaining, using, sharing, or publishing data and (2) solutions that have worked in the past. Because most of the information used by environmental scientists is currently collected by government agencies and managed in information systems, the committee focused on environmental information systems created purely or partly for public purposes. The committee could not assess the impact of database legislation, which

---

[2]Examples of market-driven changes that decreased the influence of scientists on further development include the personal computer revolution, the global expansion of the World Wide Web, and the privatization of Landsat. More powerful and less expensive computers have greatly benefited the scientific enterprise, but mass markets have resulted in a focus on parallel architectures. The benefits of such architectures have yet to be demonstrated for the large and complex multiply-connected computations that are characteristic of environmental simulations, but U.S. scientists have little option but to try to adapt to them. See NRC, 1998, *Capacity of U.S. Climate Modeling to Support Climate Assessment Activities*. National Academy Press, Washington, D.C., 65 pp. Similarly, the World Wide Web has provided a means for scientists to obtain and transfer enormous quantities of information, but participation by the general public has introduced long delays due to competition for bandwidth. An example in which privatization reduced the influence of scientists over data collection is Landsat-4 and -5. As a result of privatization, the strategy for acquiring global datasets critical to global change research was replaced by a strategy of collecting data over certain land-surface areas of interest to commercial customers. See NRC, 1997, *Bits of Power: Issues in Global Access to Scientific Data*. National Academy Press, Washington, D.C., 235 pp.

does not yet exist in the United States, even though five bills have been introduced in Congress. Similarly, database legislation in Europe (the European Union Database Directive) is too recent to have affected scientific practices.[3]

BOX 1.1
Definitions Used in This Report

The transformation of data to knowledge takes place along a continuum of processing and interpretation.

Data are numerical quantities or other factual attributes derived from observation, experiment, or calculation.

Information is a collection of data and associated explanations, interpretations, or other textual material (i.e., metadata) concerning a particular object, event, or process. **Products** are derived in a logically consistent manner from data and become input data to higher level (i.e., more processed) products. A limited number of synthesized products, here referred to as **core products**, serve a wide group of users, and have been quality controlled, calibrated, and validated according to accepted scientific standards.

Information systems are a framework for making systematic measurements and collecting, combining, and processing the resulting data into information products.

Knowledge is information organized, synthesized, or summarized to enhance comprehension, awareness, or understanding.

Understanding is the possession of a clear and complete idea of the nature, significance, or explanation of something; it is the power to render experience intelligible by ordering particulars under broad concepts.[a]

Terms used to describe the financial exploitation of public-sector data include the following:

---

[3]The key elements of the directive remain open to conflicting interpretation and controversy. See P.B. Hugenholtz, The new database right: Early case law from Europe. Ninth Annual Conference on International IP Law & Policy, Fordham University School of Law, New York, April 10-20, 2001, 13 pp.

> **Commercialization** refers to the private exploitation of government data chiefly for the purpose of financial gain. **Government commercialization** refers to government agencies charging the public for information services that were previously considered a "public good" and financed by general tax revenue.[b]
>
> **Privatization** refers to the transfer of management and control from the government to a private company or other nongovernmental organization.[c]
>
> ---
> [a] Adapted and expanded from NRC, 1995, *Preserving Scientific Data on Our Physical Universe: A New Strategy for Archiving the Nation's Scientific Information Resources.* National Academy Press, Washington, D.C., p. 10.
> [b] P.N. Weiss and P. Backlund, 1997, International information policy in conflict: Open and unrestricted access versus government commercialization, in *Borders in Cyberspace: Information Policy and the Global Information Infrastructure*, B. Kahin and C. Nesson, eds., MIT Press, Cambridge, Massachusetts, p. 300-321.
> [c] NRC, 1997, *Bits of Power: Issues in Global Access to Scientific Data.* National Academy Press, Washington, D.C., 235 pp.

THE INFORMATION TREE

This report portrays the elements of a typical environmental information system in terms of a simple analogy: an "information tree" (see Figure 1.1). The tree consists of four parts, each of which is operated by individuals working in a variety of settings (universities, government centers, commercial organizations), either independently or as members of national or international coalitions:

1. The **roots** represent data collection, in which instruments of varying sophistication are deployed and operated to collect raw data.
2. The **trunk** represents the synthesis of all available information, including new and retrospective (looking backward) raw data as well as processed information, into a limited set of core products that are useful for many purposes and serve multiple users.
3. The **branches** represent the transformation of the core products to value-added products designed to serve a specific user need.
4. The **leaves** represent the end uses of core and value-added products.

The tree also provides a means of identifying where conflicts may arise among the five stakeholder groups and where negotiations should take place.

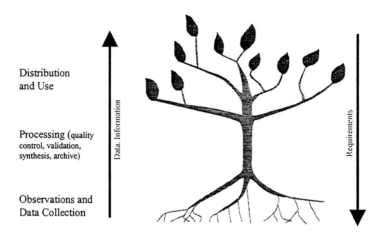

FIGURE 1.1 Schematic tree used to illustrate the elements of an environmental information system created for public purposes. Observations recorded in the roots are synthesized in the trunk and are distributed by many branches for different groups of end uses that are represented by the leaves. The information needs that these users have in common place requirements on the whole structure beneath, including the types of measurements made and the products synthesized from them. A more detailed version of the same figure is given in Figure 3.1.

ORGANIZATION OF THE REPORT

The purpose of this report is to illustrate the issues that arise in the interaction of the five environmental stakeholder groups (scientists, government agencies, private-sector organizations, policy makers, and the general public) and to outline a process by which the stakeholders can negotiate among themselves. Chapter 2 surveys the different viewpoints held by the stakeholders. The roles of each of these groups in environmental information systems are described in Chapter 3. Chapter 4 provides the economic and data policy framework for public-purpose information systems and examines the compatibility of open access to data with a competitive market. Chapter 5 illustrates the potential conflicts among stakeholders in environmental information systems created purely or partly for public purposes. Guidelines for negotiating solutions are given in Chapter 6. Finally, negotiating effectively requires that scientific and legal issues are understood by all parties; overviews of these topics are in Appendixes A and B.

2

Stakeholder Viewpoints

The five primary stakeholder groups concerned with environmental data and information are scientists, private-sector organizations, government agencies, policy makers, and the general public. Each of these stakeholder groups generates and/or uses environmental data for different purposes, according to different methodologies, and processed to different levels (see Box 1.1). This chapter describes the motivations, rewards, and mode of operation of the five environmental stakeholder groups and identifies the data policies that enhance or detract from their ability to achieve their goals.

SCIENTIST VIEWS

The goal of scientists is shared understanding among their peers. Most environmental scientists are also motivated to some degree by an ideal that shared understanding will improve the lot of all human beings and the health of the planet. Tangible rewards to scientists (e.g., tenure, salary increases, and continued research support) all derive directly from their reputation among peers for creativity, scholarship, and integrity and indirectly from the significance, productivity, and relevance to society of the field in which they work (see Appendix A).

Achieving the goal of shared understanding begins with obtaining the relevant observations, synthesizing them with information from other sources, and performing the quality control and cross-validation necessary to ensure that the resulting information product is reliable and credible. Scientific knowledge comes from challenges by other scientists that test the strength of the evidence, both during peer review in the publication process, and afterwards as the data are used in other syntheses or research purposes. Such uses require that data be available without restriction, at reasonable prices (see Box 2.1).

The nature of scientific understanding is such that it is not possible to predict reliably what data will be needed in the future. Data from unexpected sources can turn out to be very important. For example, when atmospheric carbon dioxide measurements were beginning to be taken in the late 1950s, no one realized that ice cores would provide a means of extending that climate record back in time.[1]

Science is a collective enterprise in the sense that the work of one set of specialists serves as input to the work of other specialists.[2] Nowhere is this more true than in the environmental sciences, because researchers are driven by practical reasons to collaborate.

- The environmental sciences are observational in nature and require a wide range of data from a diverse array of disciplines, taken at different temporal and spatial scales, often repeatedly over time. Controlled experiments are difficult or even impossible and existing data are often reanalyzed with new scientific objectives in mind. This puts a premium on the quality of data and surrounding information, which goes beyond the immediate purpose of data collection.
- Because nature is complex, no single instrument or observer can adequately describe the phenomena being studied. Many of these phenomena cross national borders and cannot be studied without partners in other countries.[3]

---

[1] Although ice cores have been drilled since the 1950s, collecting ice cores to address questions of climate change and global warming did not begin until the Danish-Swiss-U.S. Greenland Ice Sheet Project in 1981. See R.B. Alley, 2000, *The Two-Mile Time Machine: Ice Cores, Abrupt Climate Changes, and Our Future*. Princeton University Press, Princeton, New Jersey, 240 pp.

[2] G. Franck, 1999, Scientific communication—A vanity fair? *Science*, v. 286, p. 53-55.

[3] Environmental scientists have had a long history of international collaboration. Weather data have been exchanged around the world for nearly 100 years. The International Geophysical Year of 1957-58 led to the collection and exchange of a wide variety of earth, ocean, atmosphere, polar, and solar terrestrial data among the United States, Europe, the Soviet Union, and Japan. It also launched the World Data Center System, which archives and disseminates environmental data to the global scientific community. More recently, programs such as the World Climate Research Program and the International Geosphere

- The observations are expensive because they involve the design and deployment of specialized instruments in networks of ground stations, or on aircraft, ships, or satellites. As a result, data collectors seek to avoid duplicating the efforts of others and collaborate to save time, money, and other resources.

- Many environmental research questions require the use of all available data, past and present. Retrospective data are available through an extensive network of data centers (e.g., the National Oceanic and Atmospheric Administration's National Climatic Data Center), although scientists commonly obtain data directly from colleagues when it is easier or faster. Collections of reliable data are irreplaceable resources that are used repeatedly for purposes that are often unforeseen at the time of acquisition.[4]

The environmental sciences also differ from other branches of science in that they require access to continuous records to detect and monitor changes in the environment. Gaps in the long-term record may make some variations go undetected and others difficult to interpret.

Biosphere Program were established to document and understand the changes in the environment that are becoming apparent on a global scale.

[4]In many fields, data usage peaks immediately after data collection, then grows again as the data become part of the historical record of the condition of the environment. Such retrospective data are useful for a wide variety of scientific and public-policy purposes. Examples of the importance of long-term archives in the scientific enterprise are given in USGCRP, 1999, *Global Change Science Requirements for Long-Term Archiving*, Report from a workshop, National Center for Atmospheric Research, Boulder, Colorado, October 28-30, 1998, 78 pp.

BOX 2.1
Data Policies Relevant to Science

The environmental science community relies on policies guaranteeing full and open access to government data. Key U.S. policies governing scientific access to data include the following:

- *OMB Circular A-130, Management of Federal Information Resources* (1994).[a] Federal information is disseminated to the public on an unrestricted basis for no more than the cost of preparing a product for dissemination and distributing it to the public (incremental cost). Private contractors disseminating federal information may not impose restrictions that undercut the agencies' discharge of their information dissemination responsibilities.
- *Policy Statements on Data Management for Global Change Research* (1991).[b] Data should be provided to global change researchers on a full and open basis (i.e., without discrimination and for no more than the marginal cost of filling a specific user request).

Data policies related to international scientific programs generally specify full and open access.[c] However, some national or intergovernmental policies are more restrictive. If enforced, these could prevent scientists from publishing data used in an analysis, using the data for multiple purposes or sharing data with colleagues; yet all of these activities are needed to advance science. An example of a restrictive policy is the European Union Database Directive.[d] The directive (1) prevents the unauthorized use of substantial amounts of a database for 15 years without permission and/or payment and (2) extends intellectual property protection to facts, which are not protected under any copyright law.

The concerns of the scientific community regarding database legislation proposed in previous congresses include the following:

- Scientific facts may be covered by the legislation, and therefore are no longer widely available as a basis for further research.
- Government-produced data may receive unintentional intellectual property protection, and thus effectively be removed from the public domain.
- Traditional full and open access policies could be replaced by pay-per-use policies, which would make data less affordable (especially to scientists in developing countries) and limit the volume of data used in research projects.

- Traditional fair use exceptions (see Appendix B) would be limited, and therefore customary scientific practices (e.g., use of published information for research and education) would be inhibited.
- Excessive terms of protection (e.g., 15 years, renewable with every substantial update, in Europe) would inhibit use of new data and dramatically slow the advancement of science.[e]

Responsible scientists do not support piracy of databases. Nevertheless, changes to existing legislation concerning intellectual property must be crafted with great care. Otherwise they may upset the balances associated with copyright and fair use that have been developed both within the United States and internationally over many decades. Such balances are central to the whole enterprise of scientific research, and any change could have far-reaching and incalculable unintended consequences.

---

[a]<http://www.whitehouse.gov/omb/circulars/a130/a130.html>. In practice, most researchers obtain government data at the marginal cost of reproduction.
[b]<http://www.globalchange.gov/policies/dmwg/dmwg-gcp.html>.
[c]For example, see Earth Observation Data Policy and Europe, at <http://www.geog.ucl.ac.uk/eopole/>, and Scientific Access to Data and Information, at <http://www.codata.org/codata/data_access/index.html>.
[d]Directive 96/9/EC of the European Parliament and the Council of the European Union, on the legal protection of databases, Strasbourg, March 11, 1996.
[e]NRC, 1997, *Bits of Power: Issues in Global Access to Scientific Data*, National Academy Press, Washington, D.C., 235 pp.; NRC, 1999, *A Question of Balance: Private Rights and the Public Interest in Scientific and Technical Databases*, National Academy Press, Washington, D.C., 142 pp.

PRIVATE-SECTOR VIEWS

The goal of private-sector organizations is to sell information products or services on an ongoing basis in a commercial market. Their rewards are primarily monetary, but scientists within commercial companies may also share the same motivations and rewards of academic scientists (e.g., prestige, reputation). Likewise, many companies are motivated in part by considerations of public good, although such motivations may not be encouraged by the commercial reward system unless they can be achieved without reducing shareholder profits.

For the private sector the revenue obtained from selling the product or service must be sufficient to at least cover the costs of generating it and in the long run provide a reasonable rate of return on invested capital.[5] Competition based on satisfying customer needs at an acceptable price is the mode of operation. Depending on customer need, a commercial vendor may operate an end-to-end information system or may specialize in a particular part of the information system. Under appropriate conditions, vertical integration, in which the same enterprise exercises control over all the steps from the raw material to the final product, can provide many advantages, such as exclusive rights to all data products and their subsequent uses. Such integration is particularly important when the product is something intangible like information. In a competitive environment it is in the interest of the private sector to treat sources and techniques as commercial secrets.

The customer base for commercial data, services, and products is broad and includes entities in both the public sector (e.g., government, inter- and nongovernmental organizations, public administration, education, and research) and the private sector (e.g., agriculture, fishing, forestry, energy, natural resources, infrastructure, transportation, communications, financial, and services industries). Government agencies (especially federal agencies) are currently the primary customers[6] as well as sources of data. Indeed, it is difficult to generate an adequate revenue stream from many environmental markets unless the government pays for data collection. In the view of many private-sector organizations, once the government has developed and demonstrated the technology for collecting data, it should allow the private sector to develop applications and to market them to the public (see Box 2.2). Private-sector organizations are well placed to provide products and services that are tuned to the needs of specific paying customers because they are usually highly specialized, use sophisticated market research tools, and are responsive to the price signals provided by the market. Many of these products build upon government data (e.g., a commercial weather

---

[5] A motivation of Orbital Sciences for entering a public-private partnership with NASA was to become a player in the Earth observations industry. SOURCE: Briefing to the committee by S. Kempler, Manager, Goddard Space Flight Center Distributed Active Archive Center, on March 20, 2000.

[6] In Europe the public sector accounted for 75 percent of the market for commercial data and 59 percent of the market for commercial value-added products in 1997. SOURCE: ESYS Limited, 1997, *European EO Industry and Market: 1998 Snapshot - Final Report*, Prepared for the European Commission, Guildord, United Kingdom, 82 pp.

forecast), which usually have the advantage of being less expensive and more reliable than other sources of data, but the disadvantage of being available to competitors.

A study commissioned by the Computer & Communications Industry Association suggested the following limits on U.S. government-provided online and information activities:

- The government should exercise caution in adding specialized value to public data and information.
- The government should provide private goods only under limited circumstances, even if private-sector firms are not providing them.
- The government should provide a service online only when private provision with regulation or appropriate taxation would be less efficient.
- The government should exercise *substantial* caution in entering markets in which private-sector firms are active.
- The government should generally not aim to maximize net revenues or take actions that would reduce competition.[7]

---

[7]J.E. Stiglitz, P.R. Orszag, and J.M. Orszag, 2000, The role of government in a digital age, A report commissioned by the Computer & Communications Industry Association, 154 pp.

> **BOX 2.2**
> **Data Policies Relevant to Commercial Organizations**
>
> A number of countries are seeking to develop a commercial remote-sensing market. One means of doing so is to discourage unfair competition in the form of taxpayer-subsidized products developed by government agencies or scientists. The following U.S. policies discourage government competition with the private sector.
>
> - *OMB Circular A-76* (1999).[a] The federal government should not start or carry on any activity to provide a commercial product or service if the product or service can be procured more economically from a commercial source. The policy does not apply to products or services in the public interest.
> - *Commercial Space Act* (1998).[b] NASA should purchase space science data and space-based and airborne Earth remote-sensing data from a commercial provider to the extent possible.
> - *Land Remote Sensing Policy Act* (1992).[c] Unenhanced Landsat-4 and -5 data should be made available to federal agencies and researchers at the cost of fulfilling user requests, and unenhanced Landsat-7 data should be provided to *all* users under the same terms for noncommercial purposes. The private sector should be allowed to compete for the distribution of unenhanced data and the development of value-added services, and is encouraged to develop the remote-sensing market.
>
> Efforts to promote the development of the information industry in general are also being carried out by changing intellectual property law. For example, the rapidly growing use of information technology in fields such as publishing and communications have led governments to consider legislation that will protect databases. The primary motivation and support for database legislation comes from commercial publishers and organizations that believe that their businesses may be placed in jeopardy as a consequence of the ease (a single keystroke) with which databases to which they have proprietary rights might be transferred against their wishes. In short, their concern is with widespread electronic piracy. The European Union has already enacted database legislation (see Box 2.1 for the scientific perspective). Similar legislative bills have been introduced in the United States, but none have passed to date.
>
> ---
> [a]<http://www.whitehouse.gov/omb/circulars/a076/a076.html>.
> [b]Public Law 105-303.
> [c]Public Law 102-555.

GOVERNMENT AGENCY VIEWS

Government agencies (federal, state, local) implement public programs under the direction of policy makers. Their reward structure has two levels. For government as a whole, the reward is a populace that is better off because of a public program (e.g., the Clean Water Act) and is therefore supportive of the endeavor. For individual government agencies the reward comes from fulfilling their specific missions, the success of which is corroborated by continued funding from Congress.

The U.S. government has three major roles in the environmental information enterprise: (1) it fulfills the public need for scientific understanding by funding basic research; (2) it collects and disseminates data through a network of observing systems, agency programs, libraries, and data centers; and (3) it creates information products related to health, safety, and human welfare. The relative importance of these tasks depends on the responsible agency, which sets priorities based on its specific mission. Thus, individual government agencies are likely to have a narrower view of priorities and choices in a particular situation than the government as a whole.

Most government functions are carried out by the public sector either because of an overriding public interest in the outcome or because the potential for high risk or low payoff makes the task unattractive to the private sector. For example, federal agencies are responsible for collecting and disseminating information relevant to weather forecasting. Providing reliable data to the public requires long-term monitoring and the synthesis of current and retrospective data from around the world. The government is well placed to install and maintain the observing systems, negotiate data exchange agreements with other countries, and operate data centers that will hold the data in perpetuity.[8] As a result, the general public can obtain a wide variety of environmental data,

---

[8]For example, Congress found that "it is in the best interest of the United States to maintain a permanent, comprehensive Government archive of global Landsat and other land remote sensing data for long-term monitoring and study of the changing global environment" (Public Law 102-555). Similarly, various statutes direct NOAA to "acquire, maintain and distribute long-term databases, and to process and archive space-based data" (NASA/NOAA Memorandum of Understanding for Earth Observations Remotely Sensed Data Processing, Distribution, Archiving, and Related Science Support, July 1989).

sometimes going back 150 years and can be reasonably assured that this information will be available for future generations.

Under U.S. policy most federal government data are in the public domain and cannot be copyrighted. By making data easy and inexpensive to obtain the U.S. government seeks to promote science, create a more informed public, and foster the development of a thriving commercial information industry. Governments in other countries have similar goals, but cultural differences and economic conditions have led to the development of different data policies.[9] For example, the requirement that they recover part of their operating costs through data sales has led to the commercialization of environmental agencies in Europe and Canada. In addition, government agencies are transferring functions like environmental data collection to the private sector. As a result, data streams with economic potential (e.g., land cover, weather, geomagnetic field) are now likely to be sold rather than freely exchanged. Of greater concern, short-term private return rather than long-term social return may become the dominant criterion for selecting which observations to collect.[10] Developing countries have yet a different perspective based upon their perceptions of the potential for economic domination by foreign monopolies.[11] These different perspectives raise a potential conflict in international collaboration.

---

[9]P.N. Weiss and P. Backlund, 1997, International information policy in conflict: Open and unrestricted access versus government commercialization, in *Borders in Cyberspace: Information Policy and the Global Information Infrastructure*, B. Kahin and C. Nesson eds., MIT Press, Cambridge, Massachusetts, p. 300-321; *Toward an Integrated Data Policy Framework for Earth Observations*, Report of a workshop, Ottrott, France, July 22-24, 1996, International Space University, ISU/REP/97/1, 39 pp.

[10]A number of European government satellites have been launched with either commercial objectives (e.g., Systeme Probatoire pour l'Observation de la Terre [SPOT]) or with a mixture of commercial, scientific, and operational objectives (e.g., ENVIronment SATellite [ENVISAT]-1). See Earth Observation Data Policy and Europe, <http://www.geog.ucl.ac.uk/eopole/>.

[11]G.W. Smith, 1999, Intellectual property rights, developing countries, and TRIPs – An Overview of Issues for Consideration during the Millennium Round of Multilateral Trade Negotiations, *The Journal of World Intellectual Property*, Vol. 2, Vo. 6, November 1999, p. 969-975.

POLICY MAKER VIEWS

The goal of policy makers (elected officials and political appointees in government agencies) is to make informed judgments about what is in the best long-term interests of the communities they represent. Their rewards include a sense of satisfaction in work that benefits their community, as well as their re-election or continuation in a position of authority.

In a democracy, policy makers are accountable to the general public, which includes the other stakeholder groups. However, concerns of the general public have to be weighed against specific tradeoffs with regulation and acute local concerns. Balancing these conflicting interests for the benefit of the community as a whole is a major challenge for policy makers.

Policy makers are responsible for looking after the "big picture," such as understanding the causes of global environmental change and dealing with its consequences. They seek policies that work and are capable of evolving in view of the uncertainties that dominate the long-term projection of both economic development and environmental change. Such policies must foster (1) negotiation, when there are competing interests; (2) competition, when there is an effective stimulant; and (3) consensus on basic goals and principles. Haunting thoughtful people everywhere is the prospect of a "tragedy of the commons,"[12] in which, for lack of an effective governance mechanism, an entire resource is annihilated by the collective rational actions of all the individuals who depend on it.

The policies of a country take precedence over those of its individual government agencies. Similarly, in international settings, policies aimed at attaining a worldwide public good take precedence over those of individual countries (see Box 2.3). In such cases, governments are themselves policy makers (see "Government Agency Views" above).

---

[12]G. Hardin, 1968, The tragedy of the commons, *Science*, v. 162, p. 1243-1248.

> **BOX 2.3**
> **Intergovernmental Data Policies**
>
> Because many environmental issues transcend national borders, governments often collaborate to collect and exchange environmental information. Many intergovernmental agreements rely on full and open exchange among the parties to the agreement. Examples include the following:
>
> - *United Nations Principles on Remote Sensing* (1986).[a] Provides for access to remotely sensed Earth observations that are (1) capable of averting any phenomenon harmful to the Earth's natural environment or (2) useful to nation states affected by natural disasters or likely to be affected by impending natural disasters. The data are to be made available on a nondiscriminatory basis at reasonable prices and as promptly as possible.
> - *Organization for Economic Cooperation and Development* (1991).[b] OECD governments should strengthen their efforts to support and encourage the international science community to assess environmental risks to human health and natural ecosystems and to promote a full and open exchange of environmental data and information.
> - *Framework Convention on Climate Change* (1992).[c] All parties, taking into account their specific national priorities, shall promote and cooperate in the full, open, and prompt exchange of information related to climate change, and to the economic and social consequences of various response strategies. The full and open policy[d] was reaffirmed at the Conference of Parties in 1998.
> - *Intergovernmental Oceanographic Commission* (1993).[e] Calls for full and open sharing of datasets for all ocean programs. Data submitted for international exchange should be provided to global ocean researchers at the lowest possible cost (ideally the marginal cost) and placed in the public domain within two years of collection.
>
> In some cases intergovernmental policies of full and open exchange survived World War II and the Cold War, but are now being revised to permit member nations to commercialize their data. The classic case is the World Meteorological Organization (WMO), which has provided the framework for the exchange of meteorological data among its member nations for over a century.[f] At the 1995 WMO Congress, the previous policy of full and open exchange was replaced by a two-tiered data exchange system.[g] Tier 1 includes a minimum list of data that are required to describe and forecast accurately weather and climate, and

support WMO programs, plus any data that originating countries so designate; these data are available for full and open exchange. Tier 2 includes all remaining data; these are subject to restrictions to prevent their use for commercial purposes other than by the originating member. The change in policy was aimed at preventing private-sector entities from competing with national meteorological services in Europe, which recoup costs through sales of data and services.[h] Similar changes in IOC policy are being promoted by several European governments.[i]

---

[a] <http://www.oosa.unvienna.org/SpaceLaw/rstxt.htm>.
[b] Organization for Economic Co-operation and Development (OECD), 1994, Megascience: The OECD Forum on Global Change of Planet Earth. Paris, France, 150 pp.
[c] <http://www.unfccc.de/resource/conv/conv.html>.
[d] The international convention is "free and unrestricted." This term is fully equivalent to "full and open" used in this report.
[e] Report on Existing IOC Policy, Intergovernmental Oceanographic Commission (of UNESCO), Meeting of the Ad Hoc Working Group on Oceanographic Data Exchange Policy, IOC/INF-1144rev, UNESCO Headquarters, Paris, France, May 15-17, 2000.
[f] There are currently 185 member organizations. See <http://www.wmo.ch/indexflash.html>.
[g] WMO, 1996, *Exchanging Meteorological Data: Guidelines on Relationships in Commercial Meteorological Activities. WMO Policy and Practice*, WMO No. 837, Geneva, Switzerland, 24 pp.
[h] NRC, 1995, *On the Full and Open Exchange of Scientific Data*. National Research Council, Washington, D.C., 21 pp.; R.M. White, 1994, A cloud over weather cooperation. *Technology Review Magazine*, v. 97(4), p. 64.
[i] Oceanographic Data Exchange Policy. See <http://ioc.unesco.org/iode/>.

GENERAL PUBLIC VIEWS

The general public comprises all the members of the community, including the stakeholder groups discussed above. Because environmental information affects so many people, the community is very broad, indeed it is often global. It stands to reason that the general public hopes to increase its sense of well being through better information about its environment. Motivations are varied and range from "should I take an umbrella this morning?" to "what will the environment be like for my grandchildren?" Rewards arise from the economic benefits to them and their community from the wise use of environmental information.

The general public's stake in environmental information is enormous, but it is difficult to adequately represent their individual interests in decisions concerning information systems. Consequently, members of the general public must be represented by proxies, such as federal, state, and local government organizations, nongovernmental organizations, citizen advocacy groups, trade associations, congressional lobbyists, elected representatives, and scientific advisory committees. The mix of proxies will depend on the particular circumstances.

For this stakeholder group, access to environmental information and knowledge in a useable and convenient form is critical. Public media, such as television, newspapers, and World Wide Web sites, play a key role in delivering the information. The underlying sources of that information are and must be available on a full and open basis. Relevant policies guaranteeing full and open access to the general public include OMB Circular A-130 (see Box 2.1) and the Freedom of Information Act. Under the U.S. Freedom of Information Act agencies must make records and policy statements available for public inspection and copying.[13] European public-sector information is considerably less accessible to European citizens, in part because few countries have strong freedom of information laws.[14]

---

[13] <http://www.whitehouse.gov/omb/circulars/a110/a110.html>.

[14] *Public Sector Information: A Key Resource for Europe.* Green Paper on Public Sector Information in the Information Society, European Commission Report COM (1998) 585, Luxembourg, Belgium, 1998, 28 pp.

3

Environmental Information Systems

The goal of information systems is to establish facts and distribute information products needed by users. Some information systems are designed to fulfill a single purpose, either scientific, commercial or operational (see Box 3.1). Such systems are optimized for satisfying that specific purpose. Others capitalize on shared interests and serve a variety of user groups. The downside is that such multi-purpose systems are sometimes less flexible, always require extensive consultation and more complex decision making, and may compromise on quality for some particular purposes.

Where the environment is concerned, practical reasons frequently dictate the creation of shared information systems that serve the five stakeholder groups—scientists, government agencies, private-sector enterprises, policy makers, and the general public. Scientists are interested in obtaining as many measurements as possible and want to share the data collected for operations and decision making purposes; government agencies are seeking to stretch limited resources by building partnerships with other organizations; and private-sector organizations can more readily recover costs by exploiting technologies developed by the government or by building onto systems that have already been paid for. In addition, many environmental data can only be obtained locally but must be interpreted in a national or global context. Thus, many different nations also have an interest in shared systems. This chapter uses the analogy of a tree and, by extension, clusters of related trees to describe the attributes of each part of the information system and the roles of the different stakeholders.

BOX 3.1
Environmental Information Systems

An environmental **information system** is the framework for making systematic measurements, and collecting, combining, and processing the resulting data into information products. Information systems are necessary when (1) there are diverse or large volumes of data to manage; (2) the users are different from the data collectors; and/or (3) there are multiple sources of information that have to be integrated. Environmental information systems can be created to fulfill a single purpose or to serve multiple purposes. An example of a single-purpose environmental information system is the Tropical Rainfall Measuring Mission (TRMM) Support System, which serves scientists monitoring and studying tropical rainfall and global atmospheric circulation.[a] The TRMM Support System is used to collect data from the U.S.-Japan satellite, validate them, produce a range of data products, and distribute them to research scientists.

An example of a **shared** environmental information system is the World Weather Watch. The World Weather Watch offers up-to-the-minute worldwide weather information through member-operated observation systems and telecommunication links with nine satellites, about 10,000 land and 7,000 ship observation stations and 300 moored and drifting buoys carrying automatic weather stations.[b] The system was designed for operational meteorology, but transoceanic ships and airplanes, research scientists, the media, commercial weather services, and the general public also use this constant supply of timely data. The data management system integrates the observational networks and communications links into a coherent system, and provides a framework for managing data and products, and monitoring data and product availability and quality. The principal products are computer model results, translatable into weather maps, and a collection of validated data that provide an irreplaceable historical record of climate change that will be used by future generations of scientists in ways that cannot be foreseen.

Both the World Weather Watch and the TRMM Support System are examples of **public-purpose** environmental information systems, because they were created in the public interest. Such information systems provide a regular service for the public benefit (e.g., research, monitoring environmental changes on a long-term basis, monitoring the provisions of international treaties) as distinguished from conferring an economic advantage on a particular data producer. Other examples of environmental information systems created purely or partly for public purposes include the Global Ocean Observing System, World Meteorological Organization (WMO) Hydrology and Water Resources Programme, U.S.

> Geological Survey (USGS) streamflow network, WMO Global Atmosphere Watch, Environmental Protection Agency Clear Air Status and Trends Network, USGS National Earthquake Information Service, and Food and Agriculture Organization Current Agricultural Research Information System.
>
> ---
> [a]<http://trmm.gsfc.nasa.gov/index.html>.
> [b]<http://www.wmo.ch/index-en.html>.

THE ENVIRONMENTAL INFORMATION SYSTEM TREE

Elements of the Information System

As outlined in Chapter 1 the environmental information system tree consists of four parts: roots, trunk, branches, and leaves (see Figure 3.1). Data collection takes place in the **roots**. Because a wide variety of information is needed to address environmental issues, environmental information trees have many roots, each representing a different instrument or observing system. The organizations collecting these data are distributed around the world and include federal, state, and local government agencies, universities, private-sector companies, nongovernmental organizations, international programs, and volunteer networks (see Box 3.2). Data collection is typically the most expensive part of the information system because it is both labor intensive (particularly the collection of in situ observations) and includes remote-sensing instruments that are expensive to design, deploy, and operate.[1]

The synthesis of all sources of information gathered by the roots takes place in the **trunk** of the tree. The primary output of the information system is core products that are generated by (1) assembling the data into one location; (2) validating and cross-checking the data against other sources; and (3) synthesizing the information into products that typically have greater information content than the data from which

---

[1]For example, the national stream gauge network operated by the U.S. Geological Survey and its state and local government partners includes about 7,000 stations and costs approximately $89 million per year to operate (USGS, 1998, *A New Evaluation of the USGS Streamgaging Network: A Report to Congress*, 20 pp). For comparison, Landsat-7 cost $700 million to design, build, launch, and collect data (Will the U.S. bring down the curtain on Landsat? *Science*, v. 288, p. 2309-2311, 2000).

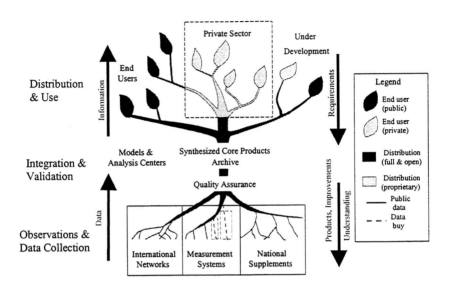

FIGURE 3.1 Tree-like structure of environmental information systems. The roots represent data collected by government agencies operating observing systems funded by an individual country (national supplements) and by government agencies, intergovernmental organizations, and research scientists participating in international networks. These observing systems are supplemented in some cases through data purchases from commercial companies (dashed lines). The data are gathered, validated, checked for quality, and synthesized into core products in the trunk by public-sector modeling and analysis centers. Distribution of these core products and the development of specialized value-added products to serve specific user communities is carried out by organizations in both the public and private sectors. Value-added products may be in the public domain (gray branches) or be proprietary (stippled branches). Both distribution avenues serve a wide range of uses (leaves), including research, commercial activities, government operations, and policy making. Users set the requirements for what core products are needed, and hence what data are collected (downward arrows), and the data are transformed into useful information and ultimately knowledge (upward arrows) through the trunk.

they were derived. The archetypal core product is the output of a large computer model that underlies a readily understandable weather map. Another type of core product is exemplified by data derived from a single rain gauge that has been accurately calibrated, fully maintained, not tampered with, and cross-checked against other sources of information to provide confidence in the results. Finally, the archive of products of the information system is itself a core product, oriented toward the anticipated needs of future generations of research scientists and other users. Each information system yields a limited number of core products based primarily on its own root system. Specification of the set of core products is a central management issue for the tree as a whole, because it determines the balance between the needs of users and the cost of meeting those needs.

Core products are created on behalf of a wide spectrum of user groups by modeling and analysis centers in research universities, government agencies and, under special circumstances, private-sector companies. Users may make substantial business, scientific, or personal investments on the assumption that the information in a core product will continue to be available. Hence, core products require a long-term commitment and a degree of care and attention that will yield a high degree of reliability and trust by a large group of users.

Although core products are useful in some sense to the broader environmental community, each distinct user group needs an add-on to the core products in the form of (1) reprocessing to make the products more accessible to that particular community; (2) combining the core products with other information special to that community; or (3) distribution and/or communication in a convenient manner or format. Every **branch** on the tree represents a value-added product or service that meets the needs of a distinct group of users. Value-added products and services are provided by government agencies, national data centers, libraries, scientists, and private-sector vendors. For example, weather information is disseminated to the public by the National Oceanic and Atmospheric Administration (NOAA) because of health and safety considerations. However, many farmers prefer to obtain weather information from private-sector vendors (e.g., Kavouras Data Transmission Network), which gather the NOAA data into a single, convenient place.[2] Scientists may repackage the same government data into weather products to facilitate teaching.

The uses of value-added products and services are represented by the **leaves** of the tree. The range of uses is broad and includes resource extraction, tourism, government operations, and research (see Box 3.3). Indeed, every human being uses some form of environmental information. The requirements for which data are collected are ultimately set by the users, who thus drive the evolution of the information system.

BOX 3.2
The Use of Volunteer Networks in Environmental Information Systems

Volunteer networks reduce the overall cost of the information system, increase the geographic coverage of reporting stations, and provide opportunities for community education. It is vital that a coherent observing system built upon this type of input provide incentives for participation. In many cases, the incentives may be little more than providing convenient access to the products that are being generated centrally and a sense of belonging to a community that is enriching human kind. This volunteer resource also requires technical support (e.g., a communication system that enables the data to be assembled reliably and inexpensively), training, observations standards, and feedback encouraging good performance, all of which must be provided by those responsible for the trunk. For example, precipitation data are being collected at over 2,500 stations in the U.S. cooperative rainfall network.[a]

Foreign governments are another kind of volunteer in the sense that (1) they are not paid directly by the U.S. government and (2) they make the decision on what observations they are willing to make and contribute to the network. Their incentives for participation are the scientific, operational, and economic benefits that arise from participating in a multi-government data-sharing arrangement. Developing countries have the added benefit of receiving training in developing and operating observing systems. Indeed, establishing volunteer networks is an effective mechanism for building science capacity.

[a]T. Karl, Director, NOAA's National Climatic Data Center, personal communication, May 14, 2001.

---

[2]Briefing to a committee-sponsored workshop by S. Goodman, director of a USGCRP-sponsored pilot project on a National Environmental Change

> **BOX 3.3**
> **Current and Potential Uses of Environmental Data**
>
> Current uses of environmental information include the following:
>
> • *Private-sector activities*, including agriculture, fishing, and forest industries; energy and natural resources industries; infrastructure, transport, and communications industries; and services industries, such as insurance, real estate, financial, news and media, software, travel, tourism, and leisure.
> • *Public-sector activities*, including public and national administration, public and operational services, education, training, and research.
>
> Potential new uses include cartography, urban planning, risk management, environment management, land registry, disaster planning, civil engineering infrastructure, and operational services.
> It is not possible to identify all distinct groups of environmental users and their needs, yet both must be taken into account in designing or updating the information system. Information on user needs can be obtained from competitive markets, where such markets exist, or from proxies such as the operators of the branches, who interact directly with the users, or community spokesgroups (see Chapter 2).
>
> SOURCE: ESYS Limited, 1997, *European EO Industry and Market: 1998 Snapshot - Final Report*, Prepared for the European Commission, Guildford, United Kingdom, 82 pp.

The Cycle for Updating Environmental Information Systems

Information systems build on existing capabilities, and participants at all levels determine which incremental changes should be made. Refining the system is a cyclic process of (1) users setting priorities for the output requirements; (2) designing a system to satisfy as many of those requirements as possible; (3) implementing the system; (4) gaining practical experience from operating the system and generating products; and (5) going back to the users and refining their statements of requirements. The cyclic nature of the system is driven by technological

---

Information System, on December 5, 2000.

advances, which enable tasks to be performed better or less expensively than before, and by new opportunities or understanding, which change user priorities over time and thus change the demands that they put on the core products. Such refinement depends upon and should contribute to a healthy infrastructure of scientific research and development that is providing new understanding for data interpretation, more capable instruments, and new useful products. On the other hand, keeping up with changes in the core products will pose a challenge to all users, including private-sector organizations, which will have to update their business plans regularly.

The trunk, roots, and branches on an environmental information tree have to work together in an efficient and cost-effective manner to serve the needs of end users. For example, the need to synthesize data and information into knowledge places three requirements on the system: (1) the tree must have multiple roots (i.e., a wide range of data taken at different temporal and spatial scales); (2) the data collection must optimize the development of core products; and (3) the products must be useful and accessible to the users. The responsibility for assuring such coherence rests ultimately with the government agencies who are funding the bulk of the operation. A significant managerial challenge is to entrain at each stage individuals with the appropriate range of experience and vision to facilitate communication between the groups responsible for ongoing implementation, system funding, needs assessment, and scientific and technical redesign. A key decision in each cycle is the specification of core products.

Conclusion. Because core products serve the needs of multiple stakeholders, a clear process by which such needs are articulated and represented in decision making is critical to the success of the information system.

4

Policy and Economic Framework for Public-Purpose Environmental Information Systems

DATA POLICY

Most environmental information systems are created and funded by the government for public purposes, including monitoring changes in the environment, research, monitoring the provisions of international agreements, and informing the public about environmental issues. Because the resulting information is built into innumerable judgments and assessments of considerable public importance, it must be highly credible. To be credible a sample of the data or data product must be independently tested and validated (see Box 4.1 and Appendix A). As a result, the core products of public-purpose information systems and the processes by which they were derived have to be open to scientific scrutiny and therefore be in the public domain. This full and open data policy satisfies the needs of scientists, government agencies, policy makers, and the general public. However, it tends to limit the enthusiasm of private-sector organizations, which rely on proprietary data for economic reasons, to be involved in public-purpose information systems.[1]

---

[1] Of course, environmental information systems for commercial purposes are being established by private-sector companies and to some extent by commercialized government agencies in other countries. The data products of these systems are created under a proprietary regime that seeks to maximize revenues by controlling the flow of information. Such a policy limits the usefulness of the data to scientists as well as to other public-sector organizations that rely on science-based products and interpretation.

> **BOX 4.1**
> **High-Quality Data and Scientific Audits**
>
> Data quality has two distinct aspects: the objective correctness of data (e.g., resolution, accuracy, consistency) and the appropriateness of data for its intended purpose.[a] High-quality data have been verified (data values are consistent with each other and meet specifications), validated (data values accord with what they represent), and certified (an appropriate authority has determined that data are correct and appropriate within specified margins for their intended use). Since data producers cannot evaluate the appropriateness of a database for an unknown user's purpose, they must provide sufficient documentation to enable the user to perform this evaluation.
>
> High-quality data are produced by both the public and private sectors. Whether commercially produced data are appropriate for scientific research or vice versa depends on the application. For example, IKONOS data are valuable for commercial purposes because of their high resolution, but their use for scientific purposes is compromised until radiometric calibration and accuracy are assessed through a scientific audit.
>
> By analogy with a financial audit, a scientific audit is performed by professionally qualified independent investigators, who check a sample of the underlying data to ensure that the product is consistent with the specifications that have been laid out. The auditor then makes a judgment about whether any discrepancies are material and by extension whether the data can be used for the purpose at hand. The result of the audit is openly available, although the input may not be.
>
> ---
> [a]J. Rothenberg, 1996, Metadata to support data quality and longevity, <http://www.computer.org/conferences/meta96/rothenberg_paper/ieee.data-quality.html>.

Nevertheless, some private-sector entities and Congress are urging government agencies to increase the involvement of the private sector in collecting and disseminating data, and creating data products (see Chapter 2, "Private-Sector Views"). Many government-developed satellite technologies are sufficiently mature that private-sector companies can profitably enter the remote-sensing industry by launching their own satellites or by developing new applications for commercial and public-sector customers. However, once established in the industry, private-sector organizations often do not want the government to compete with them by continuing to collect observations or to produce and disseminate information products. This view has been echoed by Congress in bills authorizing funding for federal agencies,[2] and in legislation forbidding agencies from competing with the private sector (see Box 2.2). Congress is seeking to create a smaller, more efficient government by transferring agency functions to the private sector (i.e., privatization; see Box 1.1) to the extent possible. *The critical question is how far can privatization go without jeopardizing the goals of a public-purpose environmental information system?*

Mechanisms for involving the private sector in government operations include (1) contracting out data collection or services, (2) establishing partnerships to collect, produce, and disseminate data and information products, and (3) privatizing the government function. In the case of contracts a commercial vendor can be required to act as an agent of the government and thus abide by the government's policy of full and open access. With public-private partnerships, a data policy must be negotiated that permits commercial objectives to be achieved while producing the credible data that the public sector needs. Because the economic and data policies of the public and private sectors are so different, successful public-private partnerships are difficult to create. Finally, under privatization the private sector gains complete control of what was the government function and sets the terms of access. In such cases the public sector becomes one of several paying customers, although certain segments (e.g., researchers) may receive data, products, or services at discounted prices.

The roles and potential conflicts among stakeholders in environmental information systems created purely or partly for public purposes are described below. Many of these conflicts derive from the well-known transition in the development of a new technology from an exploratory stage of basic research and demonstration to a mature stage

---

[2]Recent legislation noting the undesirability of government competition with the private sector includes National Aeronautics and Space Administration Authorization Act of 2000, Public Law 10-391; National Defense Authorization Act for Fiscal year 2000, Public law 106-65; National Weather Service and Related Agencies Authorization Act of 1999, Report 106-146 to accompany H.R. 1553, 106th Congress, 1st session; and Department of Interior and Related Agencies Appropriations Bill, 2001, Conference Report on H.R. 4578, House of Representatives, September 29, 2000.

of commercial application.[3] Government agencies commonly fund the exploratory stage because the investment cost and risk are high. Once the technology is well understood and reliable (the mature stage), private enterprise is more likely to develop applications and products. Core products of environmental information systems tend to be situated at just such a transition, when prototypes developed with research funds demonstrate a potential for many applications, some clearly public, but others of commercial value.

COMPATIBILITY OF OPEN ACCESS WITH A COMPETITIVE MARKET

This section presents the rationale for the following conclusions: (1) the trunk and roots of the information tree should be publicly funded to ensure that credible and dependable core products are made available on a full and open basis; (2) information derived from these core products should be distributed through value-added branches by a combination of public and commercial organizations serving different communities of end users; and (3) under certain conditions public-sector purchase of commercially available data and information services may be appropriate. The argument depends on economic considerations (i.e., the conditions under which competitive or monopolistic markets are likely to form) coupled to the particular requirements and realities of shared, public-purpose information systems.

Economic Characteristics of the Provision of Environmental Information

The provision of environmental information differs from standard production activities in two ways:

1. The marginal cost of distributing a copy of the information is typically very small, sometimes even negligible, compared to the initial cost of collecting and synthesizing the data and producing an information

---

[3]G.A. Moore, 1999, *Crossing the Chasm: Marketing and Selling High-Tech Products to Mainstream Customers*. Harperbusiness, New York, NY, 227 pp.

product. In economic terms this phenomena is referred to as declining average costs.

2. The scientific enterprise requires that basic researchers subject their findings, including a description of the data used and the models and methods of analysis, to peer review. The precise texts of articles and books may be subject to copyright, but the substance of the material must be made available to all parties willing to pay the incremental cost of distribution (e.g., the cost of a subscription or the price of a book). Thus, the production of scientific information is nonexclusionary. Furthermore, scientific information is nonrivalrous because providing the information to one party does not diminish the information available to another party. A public good, such as scientific information, is both nonexclusionary and nonrivalrous.

The first (declining average costs) is a property of the technology of producing information products, whereas the second (public-good nature of information) is largely a property of our scientific institutions. Thus, one can imagine another system in which all the products of all scientific research could be copyrighted and patented and the underlying data and analysis kept largely secret.[4] However, science would not necessarily flourish under such a system.

The declining-average-cost character of the environmental enterprise has the well-known implication that, if the organizations producing the information products are private and for-profit, there will be a tendency to create monopolies, with all their attendant inefficiencies.[5] Inefficiencies will also occur if the organizations are not-for-profit but recover all their costs from user fees (e.g., commercialized government agencies in Europe). Consequently, in the United States the primary producers of environmental information are government agencies subsidized by tax dollars and not-for-profit research organizations (e.g., research universities) funded by foundations and government grants.

---

[4]Given the ease of copying and transmitting information, keeping scientific secrets and enforcing copyright and patent laws is only partially feasible.

[5]Not all markets are created equal, and some are much more efficient than others. The ideal benchmark case is the "perfectly competitive market" in which prices are equal to marginal costs and therefore product sales occur at the efficient level. In other words, a market is efficient when all buyers willing to pay at least the marginal cost of a product are actually successful in making their purchases.

The public-good aspect of environmental information also makes it difficult, if not impossible, for commercial companies to efficiently provide the information. Thus, there are two economic arguments for relying on the public sector (government agencies and private not-for-profit entities) to collect and synthesize core information (i.e., the roots and trunk of the information system tree).

On the other hand, value-added products and services tailored to particular clients may not be characterized by declining average costs and may easily be made exclusionary (even if they remain nonrivalrous). As a result, there is a greater potential for competition in the value-added sector (the branches of the information tree), mitigating the inefficiencies of monopoly. Individual firms will retain some degree of monopoly, so one cannot expect to achieve the ideal efficiency ascribed to perfectly competitive markets. Similarly, public-sector organizations producing value-added products are also subject to inefficiencies due to "dysfunctional" incentives.[6] Society is thus forced to make case-by-case analyses of the branch activities and to search for the best institutional structures appropriate to each case. Unfortunately, the ambiguities and uncertainties inherent in these cases make them subject to ideological biases, such as a general distrust and dislike of government intervention, or conversely, a naively excessive faith in the capabilities and altruism of public servants and academics.

The Rationale for Public Funding for the Trunk and Roots

As noted above, the economic rationale for public funding of the trunk and roots of the information tree derives from the public good nature of environmental information, and the characteristic of declining average costs. The public good aspect can be summarized as follows.

The core products of the trunk are intended to help establish facts for all. It is highly undesirable for society that their use for public purposes be encumbered by intellectual property rights. The public benefits from scientists combining facts freely from many different sources to create new knowledge or understanding. Although

---

[6]C. Wolf Jr., 1988, *Markets or Governments: Choosing Between Imperfect Alternatives*. MIT Press, Cambridge, Massachusetts, 220 pp.

intellectual property rights enable individual creativity and industry to be rewarded in the marketplace, they are fundamentally a restrictive practice that distorts the exchange of information in ways that can have undesirable side effects, particularly on the conduct of environmental science.[7] For example, requiring scientists to obtain multiple permissions from possibly unknown sources before working with the data could become very burdensome, even if the fee is negligible. If the administrative burden is too great, scientists will abandon the research and the societal benefits from the potential new knowledge will not be realized.[8] Consequently, the greatest benefit from use of taxpayer resources comes from full and open access to scientific information.[9]

Recommendation. Environmental information systems that are created by the U.S. government to serve a public purpose should continue to establish facts that are accessible to all. To facilitate further distribution these facts should be made available at no more than the marginal cost of reproduction and should be useable without restriction for all purposes.

The aspect of declining average cost leads to the following observations:

- **Homogeneous product markets lead to monopolies.** In a homogeneous product market[10] the products desired by customers are identical and the only avenue for commercial competition is price. In the

---

[7]For example, see NRC, 1997, *Bits of Power: Issues in Global Access to Scientific Data.* National Academy Press, Washington, D.C., p. 132-188.

[8]The social returns to investment in basic scientific research far exceed those for the average investment dollar. See Council of Economic Advisers, *Economic Report of the President 1994*, p. 190; C.I. Jones, and J.C. Williams, 1998, Measuring the Social Return to R&D, *Quarterly Journal of Economics*, v. 113(4), p. 1119-1135.

[9]K.J. Arrow, 1962, Economic welfare and the allocation of resources for invention, in *The Rate and Direction of Inventive Activity*, Universities-National Bureau Committee for Economic Research, Princeton University Press, Princeton, New Jersey, p. 618.

[10]An example of a homogeneous product market is the market for a firm's stock shares.

long run only a single firm can operate profitably in such a market.[11] Continued profitability requires enforcement of restrictions on the redistribution of products by customers. Once a monopoly is established, prices will exceed marginal costs. As a result, the total number of sales will be lower than they would be if price equaled marginal cost. For public-purpose information, this means a reduction in social welfare.

- **The total cost of the information system is dominated by the costs of making the observations and assembling, validating, and synthesizing them into dependable, scientifically valid, well-documented products**. For the shared environmental systems under discussion measurements must be made using a variety of remotely sensed and in situ instruments, positioned both throughout the United States and in other countries, and combined with retrospective data. The infrastructure, communications, and personnel costs of participating facilities are high compared with the cost of disseminating the resulting data and information products.[12]

- **The cost of making and distributing additional copies of each core product is negligible compared to the cost of generating the master copy**. Although this situation has always been the case, copying and distributing data over the Internet or on other digital media has greatly reduced the cost of disseminating data and products relative to printed publications and older media such as microfilm. Costs of storing and accessing large volumes of data electronically continue to decline, making it feasible to copy whole collections of retrospective data, once they are in electronic format.[13]

- **It is very difficult for a commercial company to recover the cost of generating core products by selling them in a competitive market, without imposing restrictions on their re-use by customers**. The following scenario illustrates the negative impact of relaxing

---

[11]See C. Shapiro and H.R. Varian, 1999, *Information Rules*. Harvard Business School Press, Boston, Massachusetts, p. 25.

[12]An example of the high cost of data collection is given in USGS, 1998, *A New Evaluation of the USGS Streamgaging Network: A Report to Congress*, 20 pp. See also C. Shapiro and H.R. Varian, 1999, *Information Rules*. Harvard Business School Press, Boston, Massachusetts, p. 3.

[13]This is true as long as copying does not involve a transition to new technology. Migrating data to new media is generally very expensive. See NRC, 1995, *Preserving Scientific Data on Our Physical Universe*. National Academy Press, Washington, D.C., 67 pp.

enforcement restrictions on the homogenous market discussed above. Suppose that a commercial organization (the primary producer) is responsible for producing a particular core product and recovering the primary cost of production (say $100,000,000) through sales. For the organization to break even, it must set a sales price at least equal to the primary cost divided by the projected number of sales (say 10,000). In this hypothetical case the minimum sales price would be $10,000 per copy. But what if the conditions of sale did not restrict further reproduction by purchasers? In that case, the product will not be commercially viable. Without such a restriction an intermediary for a group of potential users (say 100 in number) can purchase a single copy at the primary price, reproduce that copy at negligible cost, and recover expenses ($10,000) by distributing a number of copies (say 100) among the group at the greatly reduced secondary price of only $100 per copy. Such "leakage" cuts into primary sales and reduces the total revenue of the unfortunate primary producer. Even worse, the $100 secondary price is also unrealistic, because it is vulnerable to a similar strategy pursued by another intermediary who purchases one copy at $100 but sells 10 copies at $10 each. Such competition can be expected to lower the street price until there is no possibility of the primary producer recovering even a significant fraction of the total cost of production. Under these circumstances, a lower limit to the street price is set by the marginal cost of reproduction (i.e., by the additional cost to an intermediary of making a single extra copy).

- **The social return generated from the trunk and roots of public-purpose information systems may dwarf private returns.** When a for-profit firm produces and sells data or information, even under patent and copyright protections, it may not capture in its revenues all the value that it creates for society. In this case (by definition) the total return to society exceeds the private return to the firm. Thus it is possible that the private return from a contemplative activity might be less than the cost, in which case the activity would not be undertaken by the firm, whereas the social return might exceed the cost, and hence the activity should be undertaken by society. As an extreme example, private production of a pure public good might yield a large social return but no private return whatsoever.

Conclusion. Because of the cost structure of public-purpose environmental information systems and the need for their scientific validity, the public interest is best served by funding the trunk and roots out of taxpayer resources and by providing full and open access to a set of reliable core products presenting factual information that is potentially useful to a broad range of user groups.

The Potential for Commercializing Branches

The branches provide an opportunity to develop differentiated products markets for which the prospects for rigorous competition are likely to be better. In a differentiated products market several firms produce similar but distinct products (much like the market for compact cars).[14] Determining when a competitive outcome is likely involves a two-step analysis. First, the demand and cost characteristics of each product have to be identified to determine the maximum number of products that can be profitably sold. All else being equal, a market in which demand for each product is "large" can support more vendors; conversely, as average costs increase (due to fewer sales over which the fixed costs can be spread) fewer entrants will be expected. In particular, if demand is not adequate to cover the average costs for only one firm, then no private-sector firm will be inclined to enter the market. For example, scientists alone are unlikely to constitute a viable market. Second, once the number of likely participants is identified, the nature of their interaction has to be considered. Intuitively, as the number of firms increases so does the degree of competitiveness. If a market is large enough to support only one firm and thus the sale of a single product,[15] it is likely that the monopoly firm's prices will exceed its costs and thus reduce sales relative to the competitive benchmark. Multiple firms competing in a market are also more likely to create products that meet customer demands for quality and timeliness.

---

[14]If it is cheaper for one firm to produce this set of products, compared to several firms each producing a subset, then the analysis reverts to our discussion of homogenous products (i.e., the monopoly outcome). To simplify the discussion we assume here that each firm sells a single product.

[15]If the full panoply of products were sold in this market by an equal number of firms, demand would be insufficient to cover the costs of each firm. By shrinking the set of products sold, customer demand would be shifted to the product sold by the monopoly firm, allowing it to recover costs.

The community of end users of the core products and their derivatives is diverse (see Box 3.3). It is impossible to envisage all the present and future applications of the information provided by the core products. Hence a flexible system of branches for distributing that information is essential for maximizing the societal benefit of the information system (see also next paragraph).

For some communities of end users, competing commercial enterprises that add value to the core products and other sources of information are likely to flourish. Each distinct market delineates a different branch on the information tree. For most end users, factual information in a core product has to be extracted, put in context, and combined with other facts relevant to the purpose at hand. Meeting such needs for a community of similar users can greatly increase the value of the core product itself, presenting a sales opportunity for enterprising intermediaries. In general, the diversity of customer needs and preferences within the community will enable product differentiation and individualized services tailored to those products. The larger the total demand the more competitive the market is likely to be. The specialized skills and market information that are necessary to run such a business successfully are often not available within a government agency. In such cases, private-sector operation of the corresponding branches may be appropriate.

On the other hand, public-sector operation of a branch is appropriate when the application is directly related to performance of the agency mission. In addition, the scientific research community, which has a noncommercial rewards system, may choose to organize their own branch dedicated to full and open access and paid for out of research funds. The products created by these basic research branches are likely to spawn additional specialized products as understanding of environmental processes improves or new instruments are developed. As they mature, some of these new products may open new markets for commercial application.

Conclusion. Marketing and distribution of core products and creation of value-added products is best provided by a variety of organizations, self-organized to meet the needs of different communities. Some of these value-added branches will exercise proprietary rights to products and services and operate for profit, whereas others will allow full and open access. In the latter case the source of funding will depend on circumstances, but may be public, charitable, or commercial. Yet other branches will be maintained by government agencies for public health and safety or operational purposes. Determination of the mode of operation that is most beneficial to society requires a detailed ongoing analysis of the specific circumstances for each branch, taking into account user needs for data access and standards as well as considerations of market size and differentiation.

The Potential for Purchasing Data From Commercial Entities

Provided certain conditions are met, government agencies may choose to purchase observational data from commercial entities for use in the core products. To be useful for such purposes the purchased data must be free of restrictions on use and redistribution and they must meet stringent quality standards (e.g., calibrated and adequately documented). Any compromises on documentation and openness to scientific audit (see Box 4.1) may materially detract from their value. In addition, the commercial vendor's assurances of a continued supply of data must be sufficient to justify government investment in the preparation of products that make use of them. Under these conditions, a competitive procurement based upon careful specification may lower costs below those of government operation of the same root, and hence be to the public benefit.

The principles outlined above for analyzing the competitiveness of homogeneous or differentiated products markets within which such a procurement would take place are still applicable, although the term "product" now refers to the purchased data, and the government is a customer, not a supplier. For a market that is not fully established the information necessary to apply the principles may be incomplete. However, even existing established competition among suppliers is not conclusive evidence that such competition would apply to the government procurement. Indeed, the number of suppliers may decline if the government is the sole buyer for a distinct product. This is because the government data policy and quality specifications may be so stringent that they impose substantial costs to a potential vendor, in addition to the same fixed cost of satisfying the needs of other customers. If the government were the sole buyer for this distinct product, a new

Policy and Economic Framework

homogeneous-product market would be created in which only one seller could survive. This conclusion may change, however, if the government were to procure a number of similar but distinct high-quality products. In such a differentiated products market two or more suppliers migh be able to operate successfully.

Conclusion. Purchasing full rights to data, including rights to downstream uses, from commercial entities may be an option for meeting specific observational requirements of public-purpose information systems.

REQUIREMENTS OF PUBLIC-PURPOSE ENVIRONMENTAL INFORMATION TREES

Based on the data policy and economic considerations outlined above, the committee concludes that environmental information systems created purely or partly for public purposes must meet certain requirements. Essential characteristics of the roots are:

- scientifically valid observation systems yielding quantitative data placed in the public domain;
- measurement of multiple variables at nationally and internationally distributed locations by a mixture of directed and volunteer organizations; and
- public funding with possible purchase of data from the private sector under appropriate circumstances.

Characteristics of the trunk include the following:

- systematic validation and synthesis of data into a limited selection of core information products that directly or indirectly serve all user groups;
- full and open access (provided without restriction for no more than the marginal cost of reproduction) to these products and the processes by which they are derived; and
- public funding.

Characteristics of the branches include the following:

- development based on free or low-cost access to core products from the trunk;
- value-added products for each distinct branch;
- a mixture of public domain and proprietary data policies; and
- multiple operating organizations (e.g., universities, data centers, libraries, commercial vendors, government programs).

Such clear-cut distinctions between the trunk and branches do not always exist in practice. Determining which databases are part of the trunk and which are part of the branches must be decided case by case, using the characteristics described above. Most important for the present discussion is a classification that helps maintain full and open access to data required for scientific purposes and helps promote vigorous competition where data are subject to proprietary restrictions.

Characteristics of the leaves are:

- great diversity;
- changing numbers and identities resulting from new, commonly unforeseen uses of environmental information; and
- rapidly increasing practical importance due to the growing public awareness of environmental issues.

Of course, in the real world environmental information systems cannot be described as a single tree or even a grove of trees. Rather, the roots, trunk, and branches of different information systems, some of which are operated by the private sector or commercialized government agencies, are interconnected. Data from an individual instrument may feed into the core products of several trees, which in turn contribute to the core and value-added products of other trees. Users influence in varying degrees the requirements by which several trees evolve. However, the full societal benefit will only be achieved if subsequent uses of data or products from individual trees are permitted freely.

Information systems designed to fulfill certain public objectives, such as the advancement of scientific understanding, supply products of which the use could be *but should not be* restricted. Without restrictions no private firm can recover its investment in the information system.

Policy and Economic Framework 51

With restrictions the private sector can profitably supply information products. Revenue will exceed costs, but the low level of unit sales will fail to maximize the net social benefits derived from the information system. In contrast, public agencies can subsidize creation of the system and deliver information at its marginal cost to all potential users, thereby maximizing net social benefit.

Recommendation. The practice of public funding for data collection and synthesis should continue, thereby focusing contributions of the private sector primarily on value-added distribution and specific observational systems.

5

War and Peace Among Stakeholders

Environmental information systems created for public purposes have many uses, including commercial uses (see Box 3.1). A notable example is the $500 million U.S. commercial weather industry, which uses inexpensive government-collected weather data and core products to produce commercial weather forecasts and weather derivatives.[1] The public is well served by having access to such data and services, which would not otherwise be provided by the information system. If arrangements can be made that satisfy the needs of the public sector, then commercial data are a welcome addition to the system. The same is true of government data that are restricted because of confidentiality or national security concerns or because of their commercial potential. The latter is particularly relevant in Europe. On the other hand, if restrictions prevent the data from being used in the trunk, then the data cannot be viewed as contributing to a public-sector information system.

---

[1] R.A. Guth, 2000, Japan's weather mogul to storm U.S., *Wall Street Journal*, October 30, p. B-1. Given the number of companies (more than 240) and their revenues (few millions to tens of millions each), it is likely that the $500 million figure does not include television and radio broadcasting. Weather derivatives allow businesses sensitive to the vagaries of weather to protect themselves against changes in costs and sales linked to variations in climate. These financial instruments can be designed for almost any weather variable (e.g., rain, snow, wind), although most focus on long-range (seasonal) temperature forecasts. Weather Risk Management Association, <http://www.wrma.org>. In 2000 weather-derivatives contracts with a total value of $2.5 billion were issued in the United States. PricewaterhouseCoopers, 2001, The weather risk management industry: Survey findings for November, 1997 to March, 2001. A report to the Weather Risk Management Association, <http://www.wrma.org>.

INFORMATION SYSTEMS CREATED PURELY FOR PUBLIC PURPOSES

Potential Conflicts in the Roots

The need for credibility of public-sector information products requires the input data to be available on a full and open basis or at least be subject to a scientific audit with minimal restrictions (see Box 4.1). The rigor of the scientific audit needed depends on the nature of the data and how they will be used. Restricted data that improve the efficiency of core product development but do not affect their scientific validity (e.g., base maps) may require only limited scientific scrutiny. On the other hand, when restricted data are essential to the creation of the core product, the data must be made available on a full and open basis or they cannot be used in public-sector information systems (see Example 5.1). In some cases, unacceptable restrictions on commercial data are lifted after an initial proprietary period, when the economic value has declined (see Examples 5.2 and 5.3). Such data can be an important asset to the environmental sciences, which gain a valuable new resource at a fraction of the original cost.

EXAMPLE 5.1
Hydrologic Data

Hydrologic data are used for a variety of purposes. At global and regional scales they are collected and used by scientists and government agencies to study the hydrologic cycle, determine the global water balance, and analyze climate change. At national and local scales hydrologic data are used by decision makers to monitor and allocate water resources and assess the risks of floods and droughts. Because the data have economic and military value, they are commonly sold and/or restricted by the government agency responsible for their collection or by global or regional networks that collect and exchange data.[a] For example, streamflow data are collected under the auspices of the Friend Network for use in one of its five research projects.[b] Because such streamflow data are essential for understanding and monitoring global environmental change, the United States has joined with other countries and organizations to collect historical hydrometeorological data and river basin characteristics for about 200 river basins from a range of climates throughout the world.[c] These data, which will partly duplicate the information in the Friend Network, will be freely available.

Hydrologic information is also collected and exchanged on a global basis through the World Meteorological Organization (WMO). Under WMO Resolution 25 a core set of hydrologic data are freely available for noncommercial purposes, and the remainder can be sold or restricted by the member country that collected the data.[d] Some of the core data, such as records of river flow from about 3,300 stations in 143 countries, are available through the WMO Global Runoff Data Center.[e] However, the center imposes three restrictions: (1) the amount of data that can be requested is limited; (2) users are not permitted to share the data with third parties; and (3) users must inform the center how the data will be used. Thus, the data cannot be regarded as being available on a full and open basis. The majority of hydrologic data are subject to even more restrictions and must be obtained directly from the government agency that collected them.

Lessons learned. *Limiting access to hydrologic data severely limits scientists' ability to construct or validate global or regional models of the hydrologic cycle, land-atmosphere interactions, and biogeochemical cycles. Because restricted data cannot be shared among colleagues, their use undermines the scientific practices upon which the research enterprise depends. Restrictions also create major inefficiencies in public-sector information systems because the same information must be collected by multiple organizations.*

---

[a]For example, 20 national meteorological services of Europe have joined to form ECOMET. Restrictions on data from the ECOMET network are described in their data policy at <http://www.meteo.oma.be/ECOMET/description.html>.

[b]Streamflow data are contributed from countries in Europe, South America, Africa, and Asia. In Europe alone daily streamflow data from over 5,000 gauging stations in 30 countries are archived. See <http://www.nwl.ac.uk/ih/www/research/mfriend.html>.

[c]The Model Parameter Estimation Experiment was initiated in 1999.

[d]Before WMO Resolution 25 was enacted in 1999, very little hydrologic data outside the United States was available on a full and open basis. Its passage thus had the opposite effect of WMO Resolution 40, which substantially decreased the amount of data member nations made freely available. The text of the WMO resolutions can be found at <http://www.wmo.ch/>.

[e]<http://www.wiz.uni-kassel.de/kww/irrisoft/hydro/grdc.html>.

EXAMPLE 5.2
Oil Industry Data

The U.S. oil and gas industry has spent billions of dollars to collect and produce geoscience data and information products, including geologic maps, seismic reflection profiles, and well logs. Because of corporate downsizing and a shift from domestic to foreign production, the oil industry is increasingly willing to contribute its proprietary data to facilities that operate in the public interest, such as libraries, universities, and federal agencies.[a] Although the commercial value of the data has declined, the data remain valuable for such public purposes as research and mitigation of earthquake hazards. Such information is especially valuable because areas explored decades ago are often less accessible now due to urban development (e.g., the use of seismic reflection profiles collected by Texaco, which have allowed seismologists to discover and map faults buried beneath the Los Angeles Basin[b]). Texaco provided both raw and processed data, which enabled the scientists to validate the information directly.[c] In return, the scientists agreed not to distribute the dataset without permission, except as part of normal scientific discourse.

Although oil industry data have great scientific value, transferring large quantities to public repositories can be an expensive proposition. Not only must the data be physically transferred but they must be inventoried, assessed for quality, and suitably formatted and documented. A study conducted by the American Geological Institute estimated that it would cost about $2 million per year to obtain oil industry data and operate a public archive.[d] This cost is only a tiny fraction of what the data cost to collect originally.

Lessons learned. Changing priorities in industry can lead to the transfer of valuable proprietary data to the public domain. When restrictions are lifted, data are often used in ways that were not anticipated when they were originally collected. Such additional uses greatly increase the value of the data collection, particularly to future generations. On the other hand, waiting for the commercial value to decline entails a cost to the public.

---

[a]For example, industry-collected seismic data valued at more than $1 billion reside at the U.S. Geological Survey's (USGS) National Energy Research Seismic Library, <http://energy.usgs.gov/factsheets/NERSL/nersl.html>. Shell Oil transferred its core facility to the University of Texas in 1994. A survey of the types of data the oil and gas industry would consider transferring to the public sector can be found in AGI, 1997, *National Geoscience Data Repository System:*

Phase II Final Report, American Geological Institute, Alexandria, Virginia., 127 pp.
[b]J.H. Shaw and P.M. Shearer, 1999, An elusive blind-thrust fault beneath metropolitan Los Angeles, *Science*, v. 283, p. 1516-1518.
[c]J.H. Shaw, Harvard University, personal communication, May 14, 2001.
[d]Geoscience data that might be contributed to a public repository include 100 million line miles of digital seismic data, 600,000 tapes of digital well logs, and millions of samples, paper records and analyses. See AGI, 1998, The National Geoscience Data Repository System: Promoting the preservation and accessibility of geoscience data. American Geological Association white paper, Alexandria, Virginia., 6 pp.

EXAMPLE 5.3
Systeme Probatoire pour l'Observation de la Terre (SPOT)

Commercial imagery is an important complement to the medium- and low-resolution imagery collected by the government. For example, having multiple satellites increases the chances of obtaining cloud-free views of the land surface. For many land-surface applications (e.g., agriculture, forestry, geology) data from the commercial French satellite SPOT are the best available. However, SPOT data are subject to use restrictions and the cost to users is significantly higher than Landsat-7 data.[a] Consequently, SPOT data have not been widely used for public purposes in the United States.

The use of SPOT data is likely to increase as the data become available at reduced prices through the USGS's Earth Resources Observations Systems (EROS) Data Center. The center is the backup archive for SPOT Image Corp. Under the terms of an agreement being negotiated between SPOT Image Corp. and the U.S. Geological Survey, the U.S. government and its affiliated researchers (e.g., NASA-approved investigators) will be able to obtain 700,000 SPOT images collected over North America from 1986 through 1998 for about $600 per scene, including access fees.[b] The USGS/EROS Data Center will produce and distribute the products and pay access fees to SPOT Image Corp. each time images are retrieved from the archive. SPOT Image Corp. benefits by collecting access fees without having to maintain an expensive backup archive. It is the intent of both parties to provide access to all users by 2002. When that happens, the U.S. public will benefit by gaining a valuable data resource at a fraction of the cost of building and operating a satellite.

Lessons learned. A valuable commercial resource can sometimes be made available for public purposes at a lower cost than if the government built and operated an observing system. When the data are purchased to give the public unrestricted access, the terms of the purchase should reflect the spirit of OMB A-130 (incremental cost).

[a]Imagery firms call Landsat prices unfair. *Space News*, v. 10(42), November 8, 1999. SPOT Image Corp. offers 10- and 20-m resolution images at 50 cents to $1 per square kilometer, compared with 30-m resolution Landsat-7 imagery, which is available for 2 cents per square kilometer. SPOT data are copyrighted and may not be shared freely, except among those specified in the license agreement.

[b]T.M. Holm, Chief of Data Services Branch, USGS EROS Data Center, personal communication, June, 2001; USGS adds SPOT imagery to satellite archive, *Space Daily*, March 22, 2000, <http://www.spacedaily.com/news/eo-00f.html>.

The Commercial Space Act directs NASA to purchase Earth remote-sensing data from a commercial provider to the extent possible (see Box 2.1). The intention of the legislation was to prevent government agencies from competing with private-sector organizations. At the same time, unfounded complaints about competition can stifle innovation in the government (see Example 5.4).

EXAMPLE 5.4
Demeter

Data from hyperspectral imaging instruments have many applications related to environmental changes on land and shallow marine areas. NASA's Airborne Visible/Infrared Imaging Spectrometer (AVIRIS), for instance, has been enormously successful, yielding over 500 scientific papers[a] and setting a high standard for implementation of the technology by the private sector.

Several government agencies (and their associated research scientists) and commercial companies have been considering a variety of hyperspectral missions. For example, in December 1997 the Office of Naval Research entered into a partnership with Space Technology Development Corporation (STDC) to jointly produce the Navy Earth Map Observer (NEMO) to map coastal regions and other areas of mutual interest.[b] Six months later another group of scientists and commercial partners proposed a different imaging spectrometer (Demeter) for ecological applications in NASA's Earth System Science Pathfinder competition. However, despite favorable reviews in the preliminary round of evaluations, the Demeter proposal did not advance to the next phase because NASA officials believed the mission would violate NASA and national policy to support private-sector investment in commercial space activities (the Commercial Space Act).[c] Meanwhile, STDC failed to raise sufficient funds to carry out its part of the NEMO mission,[d] with some industry representatives concluding that a commercial market for a hyperspectral sensor could not be developed at this time without more basic research. Earth Search Sciences, Inc., assumed the prime-contractor role for NEMO in late 1999, but a firm development and launch schedule does not yet exist.[e]

Lessons learned. *Innovative technologies often require government funding to reach maturity. Hyperspectral remote sensing and data products are apparently still in the research stage and will require government investment before commercialization efforts can succeed.*

[a]<http://popo.jpl.nasa.gov/html/aviris.biblios.html>.
[b]<http://www.onr.navy.mil/onr/newsrel/nr971212.htm>.
[c]Letter from NASA to S. Ustin, principal investigator of the Demeter proposal, 1998.
[d]See *Space News*, January 31, 2000, <http://www.space.com/php/spacenews/smembers/sarch/sarch00/sn0131j.php>.
[e]<http://www.earthsearch.com/technology/frame_nemo_satellite.html>.

Potential Conflicts in the Trunk

Confidence in the output of the trunk requires many creative minds to critique and verify each step of the transformation from raw data to finished products. Computer programs or algorithms with undetected errors are notorious for promulgating misinformation to the entire user community. Anything that makes scientific scrutiny more cumbersome or expensive increases the chances that errors will not be detected in a timely manner. For the shared-use information systems under discussion, such misinformation would be an intolerable outcome.

Validating data through repeated measurements or cross-checking with other independent sources turns scientific data into information (see Box 1.1). The calibration and validation steps are also necessary for detecting errors in the algorithms for processing data (see Example 5.5) or for improving the efficiency of algorithm development. Once the instrument is deployed, routine cross-checks with other instruments are essential for discovering errors, changes in instrument behavior, or scientific surprises (see Example 5.6).

EXAMPLE 5.5
Ocean TOPography EXperiment (TOPEX)

The TOPEX/Poseidon satellite was launched in 1992 to measure global sea level, monitor global ocean circulation, and improve global climate predictions.[a] Both NASA and its partner, the French Space Agency, CNES, process and distribute data to scientists. To facilitate processing, the partners agreed to exchange data and technical information on a full and open basis from the very beginning. This open data policy played a critical role in identifying a significant error early in the mission—an algorithm error that gave rise to a false global sea level rise estimate[b] with alarming implications regarding the effects of global warming. If not for the open communication among NASA and CNES engineers and science users, this mistake would have probably been resolved much later. Moreover, the spirit of collaboration resulted in many other joint projects, including the continuous improvement of algorithms, models, and processing methods.

The history of access to altimetry data from the European Space Agency (ESA) missions ERS-1 and -2 is very different. Under the assumption that the data had commercial value, ESA restricted access to members of the ERS science team. To become a member of this team scientists had to submit a formal research proposal (a no-cost proposal if not from a European Union country) to ESA requesting and justifying access to the data stream. If the proposal was accepted, the scientist had to sign a formal agreement listing all of the people by name who would be permitted access to the data. In an academic institution the turnover is so great that this restriction is difficult to satisfy. The restrictions and attendant bureaucracy, as well as technical problems with ERS-1, discouraged many scientists from attempting to use the data.[c] As a result, the ERS altimetry product has not been used as widely as the TOPEX/Poseidon product. In response to pressure from altimetry scientists and the recognition that altimetry data were not in fact commercially valuable, ESA streamlined its approval process.[d] However, ESA still requires scientists to describe their research activity and justify their need for the data.

Lessons learned. Because TOPEX/Poseidon data were available on a full and open basis, a processing error was discovered early, saving the operating agencies and science users considerable time, frustration, and expense.

---

[a]For a brief review and history, see C. Wunsch and D. Stammer, 1998, Satellite altimetry, the marine geoid, and the oceanic general circulation. *Annual Reviews of Earth and Planetary Science*, v. 26, p. 219-253.

[b]L. Fu, NASA project scientist for TOPEX/Poseidon, Jet Propulsion Laboratory, personal communication, September 1999.
[c]C. Wunsch, MIT, personal communication, June 11, 2001.
[d]Applicants are generally approved to obtain ERS data in one month; access to TOPEX/Poseidon data is usually granted within 24 hours after filling out the form on the Jet Propulsion Laboratory Web site. See <http://podaac.jpl.nasa.gov/order/order_topex.html> and <http://www-aviso.cnes.fr:8090/HTML/information/frames/general/produits_uk.html.

EXAMPLE 5.6
Antarctic Ozone Hole

A dramatic loss of ozone in the lower stratosphere over Antarctica was first noticed in the 1980s by a research group from the British Antarctic Survey (BAS) that was monitoring the atmosphere using a network of ground-based instruments.[a] The drop in ozone levels was so large that at first the scientists thought their instruments were faulty, although careful checks subsequently confirmed their measurements. Meanwhile, data from NASA's Total Ozone Mapping Spectrometer (TOMS) satellite failed to show a similar decline. The BAS results spurred NASA scientists to re-examine the TOMS data, and they found that their algorithms had been set to eliminate data with extremely low ozone levels.[b] NASA had been disregarding valid evidence for years. The reanalyzed TOMS data confirmed that the ozone loss first observed by the BAS was real and occurred over most of the Antarctic continent.

Lesson learned. Discoveries in environmental science may go undetected, sometimes for many years, simply because they are unexpected. The only safeguard is constant vigilance and scrutiny of the data and methods for analyzing them by as many scientists as possible. For this reason, environmental scientists rely on full and open access to all environmental data upon which scientific inferences are based.

---

[a]G. Carver, 1998, *The Ozone Hole Tour. Part I: The History Behind the Ozone Hole*. University of Cambridge. <http://www.atm.ch.cam.ac.uk/tour/part1.html> (April 24 2001).
[b]USGCRP, 1999, *Global Change Science Requirements for Long-Term Archiving*, Report from a workshop, National Center for Atmospheric Research, Boulder, Colorado, October 28-30, 1998, p. 12-13.

Policies of full and open access maximize the quantity and credibility of data flowing to the trunk. However, under some circumstances, restricted data that have undergone a scientific audit offer a second best source of information (see Box 4.1). For this information to be useful in scientific core products, commercial vendors must be willing to disclose enough details to establish the credibility of their sources, quality assurance, and algorithms, without permitting a competitor from replicating their commercially valuable products. As long as the data are subject to scientific audit, limited disclosure may be good enough for public purposes (see Example 5.7), though limitations of any kind reduce the opportunities for independent innovative exploitation of the data or improvement in observing technique.

EXAMPLE 5.7
Commercial High-Resolution Imagery

Global and regional land cover studies and agricultural assessments rely on medium- and low-resolution (30 m to 1 km) satellite images to provide information on vegetation conditions in ecosystems and to infer parameters such as forest biomass. Such inferences can be improved by using high-resolution images from intelligence or commercial satellites.[a] Because processing high resolution data on a global or regional basis is prohibitively expensive, scientists may prefer to use high-resolution data from specific locations (e.g., where complementary ground-based observations exist) to calibrate and validate on a sample basis inferences from lower-resolution data that has continuous coverage.

One-meter-resolution IKONOS data, which are collected by Space Imaging, Inc., are an additional confirmatory source of data for an audit of the low- to medium-resolution systems scientists depend upon. Under NASA's Scientific Data Purchase Program,[b] IKONOS data have been used to evaluate the test sites used to validate a variety of satellite sensors, including Landsat-7 Enhanced Thematic Mapper and Moderate Resolution Imaging Spectroradiometer (MODIS) and Advanced Spaceborne Thermal Emission and Reflection Radiometer (ASTER) on Terra.[c] Because the amount of data requested was small and the application had no commercial value, Space Imaging, Inc., provided complete access to the relevant data to participating researchers. Similar information is available from commercial airborne photogrammetry surveys. In this case, the choice of which technique to use to calibrate and validate the low- and medium-resolution sensors (ground observations, aircraft measurements, or high-resolution satellite measurements) is a matter of relative cost to the user.

Lessons learned. *If the information content of commercial or declassified data is well understood, the data can be very useful for scientific purposes. To exploit these data, it may be necessary to use them in limited areas because of high price or confidentiality.*

[a]H.H. Shugart, L. Bourgeau-Chavez, and E.S. Kasischke, 2000, Determination of stand properties in boreal and temperate forests using high-resolution imagery. *Forest Science*, v. 46(4), p. 478-486. Of course, high-resolution data are best interpreted within the regional context provided by medium-resolution imagery.

[b]NASA's Scientific Data Purchase program was created in response to the 1998 Commercial Space Act, which directed NASA to purchase remote-sensing data from the private sector. Thus far, commercial data have been purchased from five companies, including Space Imaging; Positive Systems, Inc.; EarthSat, Inc.; Astrovision; and EarthWatch, Inc. See <http://www.crsp.ssc.nasa.gov/>.

[c]K. Thome, University of Arizona, personal communication, May 16, 2001.

Potential Conflicts in the Branches

Many user groups want value-added products or services that make the core products more convenient to use (see Example 5.8). Such services may be provided by a wide variety of public and private-sector entities, including scientists, data centers, government project offices, nongovernmental organizations, and commercial vendors. Nevertheless, many private-sector organizations regard the creation of value-added products and services by publicly funded entities as unfair competition because they are subsidized by tax dollars (see Example 5.9).

EXAMPLE 5.8
Maps

For many environmental applications, an accurate digital elevation model or topographic map is essential for placing other data in a physical geographic context. A topographic map is itself a core product of a different information system, using sources such as classical ground surveys; the Global Positioning System; and precision photography, laser altimetry and radar from aircraft and satellites. Some of these sources were developed by the military and are classified. However, a digital elevation model derived from them might have considerable public value, yet little military significance. Similarly, high-resolution (1 m) imagery of particular scenes can be used to derive topographic map products. Because high-resolution imagery can be obtained from a number of sources, the commercial value of the image product is not fully realized unless other information (e.g., street names) is added.

To be useful for scientific purposes, the process by which the topographic map was created must be subject to scientific audit, although the raw data from which it was derived do not necessarily have to be openly available. The audit can take place in two ways: (1) by specifying the consistency and accuracy of the topographic map, then commissioning a processing system to produce it in a restricted environment (i.e., an organization that collects relevant classified or proprietary information) or (2) by statistically spot-checking this map with other information to determine its reliability. Regardless of the audit technique, subsequent uses of the map must be unrestricted. Many commercial map products can be purchased and used on that basis. Commercial vendors also have a number of options for allowing unrestricted use by reducing the commercial value of the original map product (e.g., degrading the resolution of the processed images or removing the value-added commercial elements).

Lessons learned. Core products from information systems commonly serve as input data to other core products. Great care must be taken in defining the essential requirements of an integrated product in order for a commercial vendor to provide full and open access to all data used at a reasonable price, without compromising other commercial applications from the same or similar raw data.

EXAMPLE 5.9
Weather and Climate Services

One of the most contentious current debates between the government and the private sector concerns the development of weather and climate products and services. The government has an obligation to enhance public safety and protect property, so it devotes considerable resources to collecting and processing weather data and to communicating forecasts and warnings. In some countries the costs of the weather information system are partially defrayed by selling specialized forecasts. However, a number of commercial companies take government-produced weather data and package them into value-added products, including commercial weather forecasts. The competition between government weather services in Europe and private weather forecasting companies, many of them based in the United States, led to the adoption of WMO Resolution 40. Under Resolution 40, members are permitted to restrict the commercial use of data originating in their country (see Box 2.3). One of the points of contention among members is whether scientifically motivated sharing of such data through a publicly accessible information system, thus making them available also to commercial users, constitutes a violation of Resolution 40.[a] The United

> States interpretation is that it is sufficient to label such posted data as not to be re-exported for commercial purposes.
> Whereas WMO Resolution 40 prevents private-sector organizations from competing with European governments, U.S. legislation is aimed at preventing the government from competing with the private sector (see Box 2.2). The commercial weather industry has testified to Congress on a number of occasions, calling for a reduction in services provided by NOAA's National Weather Service.[b] Some congressional committees have agreed, stating that the National Weather Service "should not directly or indirectly compete with the private sector" and directing it to confine its activities to those that "protect the lives and property of the general public."[c] Such an instruction ignores all the other public-sector users of weather data, such as climate researchers. It is not a foregone conclusion that private-sector organizations will meet public-sector needs.
>
> *Lessons learned. Legislating to prevent competition between government agencies and the private sector may not always be in the best interest of all the stakeholders. This is particularly true when there are several publicly funded activities that depend on the roots and trunk. A process for negotiating among stakeholders is needed at both the agency implementation level and the international policy level.*
>
> ---
>
> [a]P.N. Weiss and P. Backlund, 1997, International information policy in conflict: Open and unrestricted access versus government commercialization, in *Borders in Cyberspace: Information Policy and the Global Information Infrastructure*, B. Kahin and C. Nesson eds., MIT Press, Cambridge, Massachusetts, p. 300-321.
> [b]For example, see the testimony before the House of Representatives Subcommittee on Energy and Environment by Michael S. Leavitt, on behalf of the Commercial Weather Services Association on April 9, 1997, 105th Congress, 1st session; and by Joel Myers on behalf of AccuWeather, Inc., on March 25, 1998, 105th Congress, 2nd session.
> [c]Report 106-146 to accompany H.R. 1553, the National Weather Service and Related Agencies Authorization Act of 1999, 106th Congress, 1st session.

Confidential Government Data

Data that are confidential because of national security or privacy concerns pose many of the same challenges as proprietary data. To be useful in public-purpose environmental information systems, the data must undergo a scientific audit (see Example 5.10), and be produced or gridded by trusted brokers to remove the confidential elements (see Examples 5.10 and 5.11).

> **EXAMPLE 5.10**
> **Declassification of Military Data**
>
> The U.S. government has been declassifying data for many years. The effort picked up speed in 1995, when then president Clinton signed two executive orders, one calling for the declassification of documents 25 years and older and the other specifically declassifying images from the first intelligence satellite systems.[a] Declassification has led to the release of many datasets of interest to the scientific community, including geodetic data that the civilian community can now use to study global ocean floor topography (collected by the Navy Geodetic satellite, GEOSAT), data on Arctic Ocean ice cover thickness (collected by submarines) that provide information relevant to the effects of global warming, and images from current satellite systems to support scientific investigations of atmosphere-ocean heat exchange in the Arctic.[b] Declassification of images taken by U.S. intelligence satellites is being facilitated by MEDEA, a group of scientists that works closely with the intelligence community to examine and use national security data for scientific research. Their audit gives other scientists confidence in the quality and reliability of the declassified images and information derived from currently classified data.
>
> Some of the images that have been released or are being reviewed for declassification have resolutions that are comparable to the resolution of commercial imagery. For example, the best resolution of declassified CORONA reconnaissance satellite images is 2 m, compared with 1 m commercial IKONOS data, although the CORONA data were collected 40 years before IKONOS.[c] The release of GAMBIT and HEXAGON imagery, which offer submeter resolutions in small areas and wide-area coverage at coarser resolutions, is under discussion. Because of the potential impact on the commercial market for high-resolution imagery, Congress is considering authorizing the Secretary of Defense to withhold images collected by the National Imagery and Mapping Agency that could "compete with, or otherwise adversely affect, commercial operations in any existing or emerging industry, or the operation of any existing or emerging commercial market".[d] Applying this commercial filter would greatly slow the declassification of high-resolution imagery to the public and would hinder the availability of historical data. Yet, current observations are commonly far more valuable for deriving information when coupled with historical data. If data of potential commercial interest are withheld, the public will bear both the original cost of collecting the

military data and the cost of obtaining similar data from a commercial vendor.

Lessons learned. Declassified military data can be a valuable component of public-purpose information systems, particularly if a broker trusted on both sides certifies that the data are reliable. However, such data may become publicly available only if both national security and commercial concerns can be allayed. It is generally in the economic interest of private-sector organizations to restrict competition and to establish a monopoly over information sources. Such monopolies are not in the interest of the public, which deserves a good return on its investment.

---

[a]<http://www.fas.org/sgp/clinton/eo12958.html>; <http://www.fas.org/irp/offdocs/eo12951.htm>.

[b]W.H.F. Smith and D.T. Sandwell, 1994, Sea floor topography predicted from satellite altimetry and sparse shipboard bathymetry. *Journal of Geophysical Research*, v. 99, p. 21,803-21,824; <http://www.fas.org/sgp/news/1999/08/wh080299.html>.

[c]See Historical Imagery Declassification Fact Sheet, <http://www.nro.gov/corona/facts.htm>. CORONA, GAMBIT, and HEXAGON are code names for military imagery intelligence satellites.

[d]Conference report to accompany S. 1059, the National Defense Authorization Act for Fiscal Year 2000, Public law 106-65.

EXAMPLE 5.11
Hormonally Active Agents in the Environment

An emerging public health concern is the effect of hormonally and antimicrobially active agents that originate in pharmaceutical drugs and additives to animal feed. The drugs and animal-synthesized hormones pass through the digestive tracts of humans and animals, through surface and groundwater systems, and in some cases end up in the public water supply. Ingestion of hormonally active agents has been demonstrated to cause adverse reproductive and developmental effects in people and animals.[a] However, it has been difficult to define the concentrations at which specific contaminants become harmful. Existing data were collected from large rivers and streams that average contaminant levels from large populations. Collecting data at smaller scales would conceivably permit collection of contaminant information from small groups of individuals or neighborhoods. Such information would violate the rights of individuals to keep their medical records confidential.

> Similar concerns about confidentiality arose during a recent investigation of the occurrence and magnitude of environmental contaminants, including pharmaceutical contaminants, in selected regions of the United States. USGS researchers are developing a protocol for aggregating the data that permits rigorous scientific analysis while preventing disclosure of personal information. Reconciling conflicting provisions of the Freedom of Information Act with the Privacy Act[b] and other laws protecting the privacy of individuals is an issue faced by all government agencies dealing with confidential data.
>
> ***Lessons learned.*** *Some environmental problems require collaborations (e.g., between scientists and the general public) that can be formed only with the assurance that privacy will be respected. Under such circumstances it is necessary for scientists to accept less than full and open access to the underlying data.*
>
> ---
> [a]NRC, 2000, *Hormonally Active Agents in the Environment*. National Academy Press, Washington, D.C., 452 pp.
> [b]The text of the Privacy Act (5 U.S.C. 552a) can be found at <http://www.usdoj.gov/04foia/privstat.htm>.

INFORMATION SYSTEMS AND PUBLIC-PRIVATE PARTNERSHIPS

Environmental information systems are not always created solely for public purposes; many are created with a mixture of public and commercial goals. In such cases, public-private partnerships are commonly established to collect data, create products, or distribute data. A common mechanism for obtaining data through public-private partnerships is the "data buy," in which a commercial organization builds and deploys an instrument and the government agrees in advance to buy the data. Federal agencies that operate satellites are increasingly looking to data buys to cut costs, reduce financial risks, and comply with legislation prohibiting competition with the private sector. For example, NASA is currently purchasing data on ocean color (see Example 5.12), and has negotiated agreements to purchase data related to land use and land cover, climate variability, and natural hazards from five commercial remote sensing companies.[2] NASA is also considering a data purchase

---
[2]See NASA's Scientific Data Purchase program, <http://www.crsp.ssc.nasa.gov/>.

for Landsat-8, although previous attempts to privatize Landsat missions have not been successful.[3]

One of the most difficult issues to resolve in public-private partnerships is the terms of access to the data. In some cases the needs of both sectors can be met, such as when time sensitivity distinguishes the public sector and commercial markets (see Example 5.12) or when short-term commercial gain is less important than building market share in the long term (see Example 5.13). In other cases the inability to reconcile commercial and noncommercial objectives may prevent new observing systems from being built (see Example 5.14).

EXAMPLE 5.12
Sea-Viewing Wide Field-of-View Sensor (SeaWiFS)

Variations in the types and quantities of microscopic marine plants cause subtle changes in the color of the oceans. The changes in ocean color can be detected from space and used to study the ocean's role in global change and biogeochemical cycles, as well as to locate areas where fish are likely to be. The SeaWiFS instrument was launched in 1997 to meet both these scientific and commercial objectives.[a] Designed in partnership between NASA and Orbital Sciences Corp., the SeaWiFS instrument was built to scientific specifications. Orbital Sciences purchased and launched the instrument; NASA calibrated and validated the instrument and agreed to purchase data for five years. Under the terms of the data distribution policy, Orbital Sciences sells SeaWiFS data to the commercial fishing and shipping industries within 14 days of collection, then NASA obtains the rights to use the data for research purposes.[b] Five years after collection the NASA-acquired data can be used without restriction.

From the point of view of NASA researchers this joint venture has proven to be a great success.[c] The instrument is producing science-quality data, which become even more valuable scientifically as new data are collected and combined with other types of information. On the other hand, the commercial value of the data becomes negligible after the two-week proprietary period. Nonetheless, Orbital Sciences has apparently had some success in developing a fish-finding business, although it is not clear whether it will be able to recover all its costs within the five years of the guaranteed data buy.[d] Another important, although less tangible, benefit was that Orbital Sciences has established its credibility in the

---

[3]Will the U.S. bring down the curtain on Landsat? *Science*, v. 288, p. 2309-2311, 2000.

commercial remote-sensing industry. Orbital's satellite business now generates about $250 million a year.[e]

Lessons learned. Scientific and commercial objectives can be met by multi-purpose observing systems when timeliness of data access distinguishes the scientific and commercial markets. Negotiating a data policy early in the process, with participation from all the stakeholders, is essential.

---

[a]<http://seawifs.gsfc.nasa.gov/SEAWIFS/BACKGROUND/SEAWIFS_970_BROCHURE.html>.

[b]A limited number of real-time licenses are also available for (1) field experiments requiring data for ship positioning; (2) operational demonstrations to prove feasibility and usefulness; and (3) assessment of calibration, validation, and instrument performance by NASA.

[c]C.R. McClain, M.L. Cleave, G.C. Feldman, W.W. Gregg, S.B. Hooker, and N. Kuring, 1998, Science quality SeaWiFS data for global biosphere research. *Sea Technology*, v. 39, p. 10-14.

[d]Briefing to a National Research Council workshop on remote sensing and basic research: the changing environment, by E. Nicastri, EdN Consulting, on March 28, 2001.

[e]Orbital Sciences reaffirms commitment to satellites. *The Washington Post*, p. E3, March 19, 2001.

EXAMPLE 5.13
TerraServer

TerraServer is one of the world's largest online databases, providing free public access to maps and aerial photographs from the USGS, as well as images from the Russian intelligence satellites SPIN-2, at 1- to 10-m resolution.[a] A partnership of Microsoft Corporation, USGS, Russian Sovinformsputnik Interbranch Association, and other organizations, TerraServer enables users to obtain a specific 1- to 4-m-resolution image to order. The system was designed by Microsoft in 1997 as a testbed for developing advanced database technology. Microsoft operates and finances the system, and the USGS supplies digital orthophoto imagery and topographic maps. Microsoft purchased the Russian satellite data. Today, users can access over 20 terabytes of information, using commonly available computer systems and Web browsers over slow-speed communications links. Because the USGS images are in the public domain, they can be used and redistributed without restriction. The Russian data can be purchased at very reasonable prices and sample images can be downloaded at no cost.

The TerraServer has been a public success, winning a number of awards, and attracting an average of 45,000 unique users per day. It has been a success for the partners as well. The USGS is able to take advantage of cutting-edge database and software technologies to expand and improve dissemination of its maps and imagery. Microsoft is able to test its software using real data, gain the recognition associated with sponsoring a public service, and encourage a new generation to use its products.

Lessons learned. Because services may be of greater value in the marketplace than the underlying data or products, it may be in a commercial company's long-term interest to provide unrestricted or economical access to products. Their willingness to do so opens new opportunities for federal agencies to fulfill a government mandate of disseminating data to as broad a community as possible.

---

[a] <http://www.terraserver.com>.

EXAMPLE 5.14
Synthetic Aperture Radar

One of the most exciting remote-sensing technologies is synthetic aperture radar (SAR), which is used to study glaciers and ice sheets and their impact on climate; earthquake, volcano, and landslide hazards; deforestation and other ecological changes; and sea winds and surface currents. Interferometric SAR (InSAR) is used to construct precise maps of surface topography and surface change relevant to these science goals.

European, Canadian, and Japanese space agencies have flown SAR missions and are currently planning the second generation of SAR satellites.[a] All three flight agencies have memoranda of understanding with NASA that restrict the supply of SAR data to U.S. researchers. For example, under the terms of the agreement between NASA and the European Space Agency, only limited amounts of ERS-2 SAR data are available for public purposes.[b] Most of the collected data are reserved for the spacecraft owners for operational or commercial uses. However, even if more data could be acquired at affordable prices, restrictions on their use would limit their value for scientific research.

NASA has been considering launching a SAR mission for several years. Because of the high cost of such a mission and a desire to develop a commercial market for SAR data, Congress directed NASA to report on "actions the agency can undertake to support industry-led efforts to develop an operational synthetic aperture radar capability in the United States, with particular focus on NASA as a data customer."[c] NASA has thus actively sought commercial partners. Such attempts have failed so far, apparently because the commercial market is not sufficiently mature to generate a financial return within a period of time consistent with private investments.[d]

Lessons learned. When the commercial market is not mature, it may be difficult for federal agencies to find commercial partners. Under such circumstances the agency and the scientific community will have to evaluate their priorities and determine whether the data should be collected exclusively with public funding.

[a]SAR data is currently available from European (ERS-1, -2), Canadian (RADARSAT), and Japanese (Japan Earth Remote-Sensing Satellite) spacecraft, as well as from two U.S. shuttle missions (Shuttle Imaging Radar mission series and the Shuttle Radar Terrain Mapping mission). Planned missions include the European ENVIronment SATellite (ENVISAT)-1, Canadian RADARSAT II, and Japanese Advanced Land Observation Satellite spacecraft.

[b]Agreement between NASA and the European Space Agency concerning the direct reception, archiving, processing, and distribution of ERS-2 SAR data. ESA0169, August 28, 1995, 17 pp. A description of the terms of access to the European, Canadian, and Japanese spacecraft is given in NRC, 1998, *Review of NASA's Distributed Active Archive Centers*. National Academy Press, Washington, D.C., p. 109-129.

[c]Conference Report 106-379 on H.R. 2684, Departments of Veterans Affairs and Housing and Urban Development, and Independent Agencies Appropriations Act, 2000. House of Representatives, October 13, 1999.

[d]Letter from Ball Aerospace & Technologies Corp. to NASA's LightSAR Announcement of Opportunity Coordinator, May 4, 1999.

OVERALL LESSONS LEARNED

In public-purpose environmental information systems a full and open data policy is optimal for collecting and synthesizing a wide range of observations, detecting scientific surprises, and avoiding or discovering processing or calibration errors. Commercial data that are provided without restriction and at reasonable prices are a valuable addition to public-purpose information systems. Providing unrestricted access can be compatible with commercial goals, either because the commercial market will not be adversely affected by open use and publication of the data, services are of greater value than the underlying data, the priorities of the commercial vendor have changed, or because the potential long-term gain far outweighs any lost short-term profit. Restricted data can sometimes be used for public purposes, such as when a scientific audit or a trusted broker certifies the reliability of the information for the purpose at hand. However, such workarounds reduce the efficiency of information systems and have scientific and monetary costs that must be taken into account when making decisions about acquiring and using restricted data.

Privatizing government functions or creating public-private partnerships is not always the best solution for meeting the needs of all environmental stakeholder groups, particularly if the net result is a reduction of information that resides in the public domain.

THE NEED FOR A PROCESS OF NEGOTIATING AMONG STAKEHOLDERS

The requirement that the information system serve multiple uses leads to the involvement of groups of stakeholders—research scientists, private-sector organizations, government agencies, policy makers, and the general public—whose interests may not entirely coincide. Typically missing in existing management structures is a clear, identifiable process for stakeholders or their representatives to negotiate the details of solutions that optimize common interests and minimize conflicts, both at the policy level and in the details of implementation.[4] Of particular concern is the need to reconcile the requirement that sufficient high-quality data be available in the public domain (i.e., unrestricted access) with other requirements such as the need for (1) private-sector revenue; (2) protection of national security or personal privacy; or (3) demonstration of the value of investments of public funds. Solutions to these conflicts will depend on the particular circumstances of the information system at hand.

Environmental information systems frequently nucleate around informal collaborations (including volunteers) that demonstrate useful partnerships. Such collaborations have a manageable number of stakeholder groups that share enough common interests and requirements to be able to negotiate reasonable agreements. The system then evolves incrementally, limited by the ability to demonstrate real value for the costs that must be incurred and by the ability to secure necessary

---

[4]Advisory committees and workshops are good mechanisms for securing input from stakeholder groups, but they lack authority to negotiate on the stakeholders' behalf.

resources (dollars and people) to implement those improvements on an ongoing basis in order to address evolving, long-term environmental issues (see Chapter 3, "The Cycle for Updating Environmental Information Systems"). Negotiating agreements across the entire environmental enterprise is a daunting process. As the nuclei develop into long-term commitments, more formal arrangements, such as international negotiations carried out at the level of governments (e.g., Kyoto Protocol) become necessary.

On a formal basis there are two foci for negotiations, both of which are part of the cyclic process for updating the information system. One nexus is the selection of core products to be made available for public distribution, and hence of priorities for the underlying observations. The other nexus is the determination of detailed requirements for data from the individual observation systems that comprise the roots. Negotiations must address both technical issues (i.e., what data are needed to achieve stated objectives) and operational issues (i.e., who would do what and how much it would cost). The results of these negotiations provide the basis for policy decisions.

The public interest favors finding compromise solutions that are recognized by the parties concerned as reasonably satisfactory, but satisfactory agreements depend on who is at the negotiating table. For example, scientific needs (e.g., full and open access) are not always understood or taken into account because intergovernmental agreements or public-private partnerships are typically handled by government lawyers and business offices. If scientists were at the table, they would be more confident that their interests were being represented effectively. As part of the negotiations, government agencies should be prepared to provide an independent analysis of social benefits and costs using, for example, guidelines described in the following chapter. Reconciliation of the stakeholders' viewpoints is needed to produce a system that is vital and ensures environmental understanding and communal governance of the resources upon which we all depend.

Recommendation. U.S. federal agencies with responsibility for multi-purpose environmental information systems should establish a clear, visible process through which representatives of all the stakeholder groups discuss the performance and negotiate the redesign of such systems with the goal of reconciling their interests.

6

Reconciling the Views of the Stakeholders

This chapter provides guidelines for negotiating among stakeholders—research scientists, private-sector organizations, government agencies, policy makers, and the general public. Special emphasis is given to interactions between the private sector and the research community and between private-sector organizations and government agencies, where the potential for conflict is greatest.

GUIDELINES FOR INTERACTIONS BETWEEN SCIENTISTS AND PRIVATE-SECTOR ORGANIZATIONS

As illustrated in previous chapters, the involvement of private-sector organizations in collecting environmental data and creating and disseminating data products creates both problems and opportunities for environmental researchers. Environmental research scientists obtain the great majority of their data from public-purpose information systems but supplement this source by making additional measurements or purchasing commercial data. Guidelines for minimizing friction and enhancing cooperation between research scientists and private-sector organizations in public-purpose environmental information systems are given below.

Purchasing Data for Scientific Research

Commercial data providers may offer a valuable source of data to scientists (e.g., see Example 5.7), but not all environmental data collected by the commercial vendors are suitable for scientific purposes because of issues of quality, spatial or temporal coverage, price, or restrictions. Scientific concerns about purchasing commercial data include the following:

- Restrictions could prevent scientists from using data in customary ways, such as sharing data with colleagues, using them for multiple purposes, or publishing them in scientific journals.
- Research proposals that include the cost of purchasing data (especially projects that require large volumes of data) are likely to be more expensive and thus less competitive in the proposal process.
- Mechanisms for the scientific community to convey their collective needs to commercial vendors are currently insufficient.
- Commercial data may not meet scientific specifications.
- Many commercial instruments are tasked to collect data only in specific areas, which prevents assembly of global datasets.
- Documentation of the quality of the data and the methods by which they were acquired may be inadequate. Complete documentation would include details of the design of the instruments used, the results of calibration experiments, the way instruments were deployed and their sampling, random and systematic measurement errors determined from comparisons with other data, the algorithms used to process the data, and a log of exceptional circumstances surrounding the measurements themselves.
- There is usually no provision for long-term archival of commercial data, which leads to gaps in the long-term record of the environment and hinders the research of future generations of scientists.

On the one hand, scientists encountering barriers such as use restrictions commonly abandon a particular line of research, rather than invest resources to work around the barrier. The impact of such missed opportunities is difficult to assess. On the other hand, scientists are opportunists who will work with whatever data are available. If the data are intended to be used for just a short-term research project, restrictions on subsequent uses may be acceptable. If the purchased data will also be contributed to an archive used by a broader community, then restrictions on other uses may undermine the long-term interests of science as a whole.

Conclusion. Scientific practices require (1) full and open access to data and core products and (2) peer-reviewed publication. In addition, global change research requires the compilation of high-quality, long-term, global databases that are suitable for a wide variety of scientific purposes. Private-sector entities that supply data to public-purpose information systems must conform to these practices.

Scientists can obtain commercial data either by purchasing them directly from commercial vendors or by participating as investigators in a government-sponsored data purchase program.[1] Scientists using research grants to buy data have a number of possible workarounds for reducing their costs.

- Use data offered at an educational discount.
- Purchase small quantities of data and tailor the research accordingly (e.g., Example 5.7).
- Redefine specifications to avoid commercial elements, such as by using only spectral bands from high-resolution instruments that are of little or no interest to commercial customers.
- Delay purchasing data until the commercial value declines (e.g., Example 5.3).[2]
- Substitute something for the raw data, such as data that have been manipulated to remove commercial (or confidential) elements (e.g., Examples 5.10 and 5.11).

---

[1] For example, under NASA's $50 million Scientific Data Purchase program, NASA-approved researchers can obtain access to data already collected by certain commercial vendors or they can task selected satellites or aircraft (e.g., IKONOS) to make new observations in specific locations. The terms of access are negotiated by NASA with input from the research community on issues of data quality, science relevance, data usability, likely breadth of use, levels of collaboration, and data rights.

[2] Some applications, such as monitoring natural disasters, require near real-time access to data. In such cases, scientists may have no choice but to use restricted data. See J.E. Janowiak, R.J. Joyce, and Y. Yarosh, 2001, A real-time global half-hour pixel-resolution infrared dataset and its applications, *Bulletin of the American Meteorological Society*, v. 82(2), p. 205-218.

If the quantity of data needed is small or if the commercial elements have been avoided or removed, the vendor may permit full and open access because sharing and publishing the data in a customary manner would not harm the commercial market (e.g., Example 5.2). If the commercial vendor will not permit full and open access, the researcher must switch to other sources of data or help justify spending taxpayer resources on a new government-sponsored observing system.

Conclusion. It is sometimes possible to work around restrictions but doing so is inefficient and requires scientists to modify their research objectives.

A scientific audit conducted by respected and trusted scientists at the time of data delivery might also be an acceptable compromise for ensuring adequate documentation. Disclosure of the scientific quality of such data and the circumstances surrounding their acquisition might provide an advantage to competing vendors. Yet, if such metadata is not gathered and carefully recorded at the time of observation, it is usually impossible to reconstruct the information later. In that case, even if the data meet minimal performance requirements, their value for future use is greatly diminished. The auditors would (1) ensure that complete and satisfactory documentation exists; (2) publish a summary of its scientifically significant conclusions; and (3) obtain assurances that complete documentation would be published as soon as its commercial significance has decreased.

Conclusion. Confidential scientific audits of commercial data followed later by full disclosure may be a valuable tool in assuring data quality.

Purchasing Value-Added Products and Services

Value-added products and services are commonly created by private-sector organizations. Commercial value-added products that meet the needs of the scientific community are welcomed and used by most researchers. If the product is too expensive for research budgets, scientists will create their own from the same openly accessible sources of data. Such competition with the private sector is fair, as long as funding for developing the product is obtained through the scientific proposal process and is evaluated against other research activities.

Conclusion. If a value-added product is justified scientifically but is too expensive to purchase from a commercial vendor, then scientists are justified in creating the product, as long as all costs are paid from peer-reviewed research budgets.

GUIDELINES FOR INTERACTIONS BETWEEN GOVERNMENT AGENCIES AND PRIVATE-SECTOR ORGANIZATIONS

At the instigation of Congress, government agencies are promoting the development of a commercial remote-sensing industry. In the initial stages agencies are purchasing data and services from commercial vendors, either by contracting with private-sector organizations or by forming public-private partnerships. Eventually some of these products and services will be privatized entirely. As noted in Chapter 4, private firms and market mechanisms should be considered when they advance the interests of society.

Guidelines for determining the respective roles of the government and private sector have been proposed elsewhere. For example, an industry study divided activities into three categories: those that are clearly public and should be undertaken by the government; those that are in the domain of the private sector; and those that have both public and private benefits, which the government should undertake only after careful consideration.[3] Environmental information systems generally fall into the latter category.

This section provides criteria for government and private-sector interactions concerning the roots and branches of public-purpose information systems. The criteria are meant to ensure that public-sector needs continue to be met when the private sector provides the data or

---

[3] J.E. Stiglitz, P.R. Orszag, and J.M. Orszag, 2000, The role of government in a digital age, A report commissioned by the Computer & Communications Industry Association, 154 pp. The goals of these criteria, which are meant to maximize opportunities for the private sector, are somewhat different from those discussed in this report, which aim at maximizing the public good. Of course, these views are not necessarily mutually exclusive.

service. The resulting arrangements are likely to be complex, particularly if the information system serves multiple objectives.[4]

Purchasing Data and Public-Private Partnerships

Under some circumstances data collected by commercial entities or by public-private partnerships may be suitable for use in public-purpose information systems. Criteria for government agencies purchasing data for public-purpose information systems include the following:

- The data must fulfill public-sector needs. For example, data intended for scientific purposes must fulfill both immediate research objectives and long-term scientific goals. This in turn requires that the government acquire all the data rights (i.e., full and open access) within a specified period of time.
- The data must be of suitable quality and have undergone credible calibration and validation techniques to assure they meet that quality. If the objectives include contributing to future research needs, full documentation is required (see "Purchasing Data for Scientific Research" above). When the data are initially available for confidential scientific audit, the vendor must make a commitment to publish the full record of that audit when commercial reasons for confidentiality are no longer applicable.
- Because it takes several years to fully develop data products, there must be a reasonable prospect of a long-term supply of data.

---

[4]For example, legislation governing data policy for Landsat-7 seeks to achieve multiple goals: (1) ensure that unenhanced data are available to all users at the cost of fulfilling user requests; (2) ensure timely and dependable delivery of unenhanced data to the full spectrum of civilian, national security, commercial, and foreign users and the National Satellite Land Remote Sensing Data Archive; (3) ensure that the United States retains ownership of all unenhanced data generated by Landsat-7; (4) support the development of the commercial market for remote sensing data; (5) ensure that the provision of commercial value-added services based on remote-sensing data remains exclusively the function of the private sector; and (6) to the extent possible, ensure that the data distribution system for Landsat-7 is compatible with the Earth Observing System Data and Information System (Land Remote Sensing Policy Act of 1992, Public Law 102-555).

These criteria also hold true for public-private partnerships, which are becoming an increasingly common mechanism for collecting environmental data. Such partnerships differ from data purchase agreements in that there is a quid pro quo, and therefore more incentive to compromise. Although some public-private partnerships have been a success (e.g., Examples 5.12 and 5.13), others have not (e.g., Landsat-4 and -5) or have not passed the negotiations stage (e.g., Example 5.15). Before entering negotiations with potential private-sector partners, government agencies should produce a data plan to ensure that their mission and long-term strategy are fulfilled.

Commercializing Government Data

Commercialization refers to the financial exploitation of government data (see Box 1.1). U.S. information policy (particularly OMB Circulars A-130 and A-110; see Box 2.1) encourages such exploitation by stipulating nondiscriminatory access at the marginal or incremental cost of reproduction. Because the U.S. government does not hold intellectual property rights (see Appendix B), commercial exploitation of government data can coexist with public-sector uses, such as scientific research. Such open data policies are partially responsible for the success of the U.S. information industry[5] and research enterprise.[6]

Conclusion. The commercialization of U.S. government data maximizes the use and thus the value of data in both the public and private sectors.

---

[5]According to a European Commission report, "Since the Freedom on Information Act was enacted in 1966, the US government has pursued a very active policy of both access to and commercial exploitation of public sector information. This has greatly stimulated the development of the US information industry." See *Public Sector Information: A Key Resource for Europe*. Green Paper on Public Sector Information in the Information Society, European Commission Report COM(1998)585, Luxembourg, Belgium, 1998, 28 pp.

[6]NRC, 1997, *Bits of Power: Issues in Global Access to Scientific Data*. National Academy Press, Washington, D.C., p. 17.

In contrast to the U.S. approach government agencies in other countries exert copyright, database protections, and other forms of intellectual property in order to control access to data (see Chapter 2), thereby acting like monopoly suppliers and limiting the extent to which government-collected data can be used.

These different approaches to data access are commonly a point of contention in international collaborations (e.g., Examples 5.1 and 5.5). Yet, such collaborations are essential for addressing global or regional environmental problems. Full and open access has been the norm in international collaborations for the following reasons:

- It is necessary for establishing confidence in the data, both for current and future uses.
- It is difficult to predict what information and in what amount will be needed to address the problem.
- It facilitates the integration of global datasets, as well as the widespread application of knowledge gained about environmental processes in a particular region.
- Creating multiple copies of data and metadata through open sharing increases their chance of long-term survival.

Even if scientists and government agencies in wealthy countries can afford to buy data and the associated data rights from commercialized government agencies, it is unlikely that developing countries will be able to do so. If developing countries are excluded from long-term programs, it will not be possible to obtain the comprehensive coverage and range of expertise needed to address many global environmental problems.

Conclusion. A data policy of full and open access that provides for unrestricted uses maximizes the benefits of international collaboration and the social benefits of the scientific endeavor.

Privatizing Government Functions

Determining which functions should be public and which ones should be private is the object of a long-standing debate. It is commonly in the public's interest[7] to transfer government functions to the private sector, provided that competition leads to better products and services at lower prices, and public resources can be directed to other priorities.[8] On the other hand, where the environment is concerned, privatization may have some pitfalls, including

- a likelihood that the market structure will evolve toward a monopoly, which would increase the cost of data to all users and probably diminish responsiveness to user needs;
- a commercial market may not exist and thus measurements important for some public uses (e.g., research, operations) may not be made at all (Example 5.4); and
- a commercial market may lose viability and thus measurements may be discontinued, creating gaps in the long-term environmental record.

Government attempts to stimulate commercial markets for Earth observation data in Europe have not been entirely successful. A European market study indicated that total revenues for commercial Earth observation data grew only 6 percent from 1994 to 1997.[9] The same study indicated an apparent tendency toward industry concentration. In 1997, 13 companies captured 50 percent of the market, down from 16 companies in 1994. Little had changed by 1999,[10] and a United Kingdom Parliament committee noted, "Despite more than a decade trying to stimulate commercial markets for Earth observation

---

[7]For purposes of this report, public-sector interests include scientific research on the environment, health and safety issues, and government operations.

[8]A general discussion of the economics of privatization can be found in J. Vickers and G. Yarrow, 1988, *Privatization: An Economic Analysis*, MIT Press, Cambridge, Massachusetts, 454 pp.; and D.M. Newbery, 2000, *Privatization, Restructuring, and Regulation of Network Utilities*, MIT Press, Cambridge, Massachusetts, 466 pp.

[9]ESYS Limited, 1997, *European EO Industry and Market: 1998 Snapshot - Final Report*, Prepared for the European Commission, Guildord, United Kingdom, 82 pp.

[10]Presentation to a European Commission workshop, *Has EO found its customers?*, by S. Howes, ESYS Limited, on April 21-22, 1999. See <http://co conuds.nlr.nl/workshop_4-99/ceo_workshop.htm..>

data, provided at public expense, it is universally accepted that the take-up has been unsatisfactory.... Further EO expenditure at current levels must be driven by more than an expression of general but unsubstantiated hope that commercial markets will be generated. It must also be accepted that there are good public policy reasons to gather EO data which cannot be expected to generate a commercial return."[11] Similarly, the U.S. Congress concluded that "commercialization of the Landsat program cannot be achieved within the foreseeable future."[12]

This situation may change as the market matures. According to Geoffrey Moore, when a new technology is introduced, it follows a predictable path of marketplace adoption.[13] First, a small group of visionaries who like new things and are looking for breakthroughs buys the product. The early adopters, who take on high-risk products in hopes of high rewards, soon follow. The pragmatists, who make up the bulk of the market, enter only when the products are well established and well supported. In the case of environmental information systems, scientists are the visionaries, government agencies are the early adopters, and private-sector organizations are the pragmatists. Privatization is possible when and if the chasm from early adoption to the mainstream market is crossed.

Privatizing Branches

Private-sector organizations may be better positioned than government agencies to identify potential applications. If a viable commercial market for the value-added product exists, it may be in the public interest to encourage private-sector organizations to create that product. Government resources could then be devoted to developing products that benefit broad and diffuse groups of users or are considered too risky for the private sector to undertake. On the other hand, if suitable commercial products are too expensive (in terms of price or restrictions), it may be appropriate for the public sector to provide the value-added product or service. Criteria for the government to

---

[11]United Kingdom House of Commons, Trade and Industry Committee, Tenth Report. <http://www.parliament.the-stationery-office.co.uk/pa/cm199900/cmselect/cmtrdind/335/33502.htm>.

[12]Public Law 102-555.

[13]G.A. Moore, *Crossing the Chasm: Marketing and Selling High-Tech Products to Mainstream Customers.* Harperbusiness, New York, NY, 227 pp.

discontinue an existing product line in favor of its production by private entities include the following:

- There are no overriding considerations, such as health and safety, that require continuing government control.
- Government-funded functions (e.g., scientific research, education, government operations) would not suffer greatly from restricted access to the product.
- Private-sector organizations are interested in taking over the functions of the government-funded branch.
- The research and development underpinning the application is mature (i.e., there are no significant uncertainties surrounding the interpretation of the available data for the purpose at hand).
- A demonstrable market that supports vigorous competition exists or is at least plausible, thereby reducing the possibility of a private monopoly. This is particularly important when the government and its affiliates would be the major customers.
- Existing users would not be harmed significantly if product availability were interrupted. For example, gaps in the long-term climate record may prevent detection of rapid temperature changes, thereby hindering scientific research and environmental policy making.

To avoid creating a monopolized market it is important that the following three conditions be satisfied before proceeding with privatization: (1) the prospective products are substantially differentiated; (2) it makes financial sense for separate firms to offer these distinct products; and (3) the affected market is of sufficient size to support at least two but preferably three or more firms (see Chapter 4). If a market is too immature to identify this information with any reasonable certainty, privatization may be premature.

Government agencies should not feel compelled to discontinue a service that is to the public benefit simply because a commercial vendor chooses to duplicate it. Similarly, government agencies should be permitted to replicate a privatized service if regeneration can be done at an incremental cost smaller than that of purchasing full rights from a private vendor or if questions about data quality or continuity arise. Information vital to the public interest should not be "captured" by the private sector, which has economic reasons for controlling access.

Privatizing Roots

Criteria for government agencies to discontinue data collection and purchase data from private-sector organizations include the following:

- A commercial capability for supplying the necessary data exists.
- The private sector is likely to provide a stable, long-term information supply.
- There is an effective process for conveying scientific or operational requirements to the private sector.
- The content and conditions of access to datasets (in particular, full and open access) would fulfill public-sector needs.
- There is an established process for ensuring quality assurance and quality control of the commercial data.
- A substantial commercial market for the data exists that would not be compromised by the full and open access provided to the government and could reduce costs to the government.

If all of the above conditions are fulfilled for public-purpose information systems, it may well be in the public interest for the government to privatize data collection. In such a case, continued provision of data by a government agency will likely discourage private-sector organizations from building quality services that better meet the needs of the public.

On the other hand, privatization is not without risk because it involves discontinuing government functions with proven value in favor of private-sector services for which benefits may never accrue. The risks are greatest in data collection because of the potential for gaps in the long-term record of environmental change. Nevertheless, under certain conditions, the collection of data and/or generation of data products can be transferred from the government to the private sector. Care must be taken to ensure that high-quality measurements and products needed by the public sector continue to be made, that data will continue to be made available on a full and open basis (i.e., without restriction and for no more than the cost of reproduction), and that the commercial vendors operate in a competitive market.

Decisions concerning which functions should be public and which should be private must be made case by case, using criteria such as those outlined above. Most decisions will involve the transfer of government functions to the private sector but some will concern the reverse situation. For example, if previous privatization efforts have led to a costly monopoly, a decline in data quality, or gaps in the long-term record, then re-entry by a government agency may be desirable. Of course, such decisions must be re-evaluated as circumstances change. Policy makers cannot expect to be able to write a single rule that applies to all cases or for all time.

Recommendation. Before transferring government data collection and product development to private-sector organizations, the U.S. government should ensure that the following conditions will be satisfied: (1) avoidance of market conditions that will give any firms significant monopoly power; (2) preservation of full and open access to core data products; (3) assurance that a supply of high-quality information will continue to exist; and (4) minimized disruption to ongoing uses and applications.

Appendixes

Appendix A
Scientific Practices

To examine how any change in data policy or database legislation might affect research in environmental science it is important to understand how the scientific research community uses data.

The basic principles that govern this aspect of modern scientific research have evolved so as to promote and reward creativity among individual researchers in a context of a professional ethics code aimed at preserving the credibility and integrity of the scientific process. Two essential mechanisms are employed to achieve this goal: (1) publication of data and results, including all reasoning and data processing steps; and (2) peer review of all information published, in a way that mitigates potential conflicts of interest, and fosters open debate of issues on which consensus has yet to be reached. These mechanisms have been recognized as key elements in the extraordinary advances of the scientific endeavor over the past half-century.[1] As we shall see, this has a profound impact on the use of data and databases in the sciences, and especially in the environmental sciences.

The creativity of individual researchers is the wellspring from which the whole scientific enterprise flows. Donald E. Stokes observed that "it was the American research universities . . . that converted original scientific research into an economically viable professional career." Yet, most academic scientists choose to work in university research environments not for the salary (they can often make far more in the private sector) but for the intellectual stimulation that comes from scientific analysis, interactions with students and peers, and the freedom to work on problems of their own choosing. Various traditions have sprung up to foster and reward such creativity. Professional recognition

---

[1] D.E. Stokes, 1997, *Pasteur's Quadrant: Basic Science and Technological Innovation,* Brookings Institution Press, Washington, D.C., 180 pp.

and rewards for scientists (e.g., promotion and tenure in academia, salary increases, continued research support, and professional medals and honors) all derive ultimately from the reputation of the individual scientist among peers. That reputation is based primarily on the evidence contained in published work. Publication—in the etymological sense of "making open to the public"—is the primary means by which the evidence is established and the reputation of a scientist is thereby created. Of special value is the initial publication of any important contribution to knowledge, be it in the form of newly acquired data, fresh analysis and interpretation of such data, or innovative theory suggested by the observations. Scholarship is also recognized in other works, such as summaries of existing knowledge in a form that makes it more accessible to one's peers.

To achieve peer recognition scientists want their work to be read to the point where they are commonly willing not only to defray the cost of publication through page charges but also to sign over intellectual property rights (and potential profits) to publishers. It is in the interest of scientific authors to facilitate use and re-use of their data and results by peers and to contribute this material to community databases if doing so results in increased recognition.[2] Scientific progress depends critically on such practices, because rarely if ever does a major scientific advance proceed from a sudden flash of insight ("Eureka!"), even if that may on occasion appear to be the case. In truth the insight usually derives from an entire body of knowledge and a vast catalog of facts that has been assimilated by scientists. This is particularly true of the environmental sciences, in which the object of study is typically complex in the modern mathematical sense and advances proceed through patient analysis and re-analysis of growing volumes of observations, resulting in increasingly reliable models that are ever closer to the observable reality.

Because the publication process is so important to science—as it is to most creative activities—it must be afforded some degree of protection. Good science requires an intellectual property regime that balances individual rights of ownership against public rights of access to and reuse of fundamental data. In science, intellectual property protection is typically provided through copyright, as well as through a strong code of professional ethics, that incorporates the following broad elements:

---

[2]Of course, in some instances the advancement of science takes precedence over personal recognition.

Appendix A

- Respect copyright protection by adhering to the limits dictated by the fair use exception (see Appendix B) and by avoiding plagiarism.
- Acknowledge and document relevant sources of information and prior published work.
- Adhere carefully and honestly to accepted standards of record keeping and reporting of results.

The effectiveness of this mechanism depends on several levels of enforcement:

- The implicit social contract between scientists and publishers is enforceable through legal means, primarily by application of copyright law and the fair use doctrine.
- Peer pressure exercised in the course of peer review of proposals or articles leads to a loss of professional reputation and associated professional rewards (e.g., for improper attribution of previous results).
- Egregious misconduct (e.g., plagiarism, data falsification) will normally trigger an investigation by the parent institution or a professional society.

Overall, the system appears to operate well, providing both scientists and publishers with adequate protection of their intellectual property rights.

The credibility of the scientific process and the knowledge that is ultimately derived is assured by a number of mutually reinforcing activities:

- use of accepted standards of evidence and methods of inference based upon established knowledge;
- peer review prior to publication;
- independent replication of results;
- critical examination of published work to distinguish established fact from controversial hypotheses and to expose inconsistencies or ambiguities in current theory or evidence; and
- the exercise of integrity and good judgment by participating scientists.

The infrastructure of scientific publication and community criticism is particularly important in scientific assessments of the state of knowledge in particular areas. An example is the assessment of global warming, which is carried out every five years by hundreds of scientists from around the world under the auspices of the Intergovernmental Panel on Climate Change. Such assessments require that participating scientists (and reviewers) obtain data from a variety of sources, integrate them with other types of data, and use them in ways that were not envisioned when they were collected. The credibility of the results depends on adherence to established scientific practices. Scientists are trained to be skeptical of established dogma, and complete unanimity is thus unlikely. Nevertheless, ongoing review of the results and of subsequent work helps establish when a broad consensus exists, when competing theories remain controversial, and when evidence is speculative.

Appendix B
Intellectual Property Rights to Data

Intellectual property law seeks to give authors enough control over their work to motivate them to create and disseminate, while limiting their control so that society as a whole benefits from access to that work.[1] The types of intellectual property most relevant to public-purpose information systems discussed in this report are copyright, database protection and contract.

Copyright and Fair Use

Copyright law protects the creative elements of a compilation—the selection, coordination, and arrangement of the information—although not the facts themselves. For example, the yellow pages are protected by copyright because such features as the organization of information, use of boxes, and color required thought and creativity. On the other hand, the white pages are a simple alphabetical listing, which is *not* protected by copyright. Most datasets used by scientists either fall under copyright law or are in the public domain and are available to all without conditions. (By law the U.S. federal government cannot copyright databases.) Scientists can generally use copyrighted material because of a fair use exception in the United States or similar exemptions in Europe (see Box B.1).

---

[1] Intellectual property laws relevant to the environmental science enterprise are discussed extensively in NRC, 2000, *The Digital Dilemma: Intellectual Property in the Information Age*. National Academy Press, Washington, D.C., 340 pp.; and NRC, 1997, *Bits of Power: Issues in Global Access to Scientific Data*. National Academy Press, Washington, D.C., 235 pp.

> **BOX B.1**
> **Fair Use Exceptions in U.S. Copyright Law**
>
> Fair use is a bedrock principle that reconciles the Copyright Act's grant of exclusive rights to authors and the First Amendment's constitutional guarantee of free speech. Under copyright certain public purposes including "criticism, comment, news reporting, teaching, scholarship, or research" are permitted.
>
> U.S. courts consider four factors when determining whether the fair use exception is allowable:
>
> 1. the purpose and character of the use, including whether such use is commercial or is for nonprofit educational purposes;
> 2. the nature of the copyrighted work;
> 3. the amount and substantiality of the portion used in relation to the copyrighted work as a whole; and
> 4. the effect of the use on the potential market for or value of the copyrighted work (most heavily weighted factor).
>
> Decades of court interpretations define what is meant by fair use today. A fair use exception is most likely to be granted under the following conditions:
>
> - the use is for noncommercial purposes;
> - the original work was inexpensive to produce and/or distribute;
> - a relatively small amount of the original work is used;
> - the portion used is transformed, not merely copied; and
> - the economic impact is insignificant.
>
> SOURCE: K.D. Crews, 1995, Fair use: Overview and meaning for higher education, in The Trustees of California State University, *A Copyright Handbook*, p. 37-41; A.P. Lutzker, 1996, Fair use in transition: Reconciling fair use to the world of electronic communications. Intellectual Property Law Institute, State Bar of Texas, Austin, January 1996, p. G1-G4.

A primary protection against violations of copyright has been the cost and nuisance of physically copying protected material. However, with the increasing use of digital information, the growth of computer networks, and the creation of the World Wide Web individuals can now copy and distribute publications and large amounts of data at little cost or effort. As a result, existing intellectual property laws are being altered

Appendix B 97

and new laws are being considered to legislate new rights and limitations on those rights in the digital environment.[2] Among the most significant new legislation enacted or being considered is database protection.

Database Protection

Databases produced by "sweat of the brow" (i.e., created with money, effort, or labor but without creativity) are not protected by copyright. In Europe databases are protected by a new form of intellectual property protection through the European Union (EU) Database Directive (see Box 2.1 and 2.2). The EU Database Directive provides 15 years of protection for the contents of the database (including facts) and each significant update. Use of substantial parts of a database and value-added uses of any kind are prohibited without permission and/or payment to the database owner. European countries are permitted to designate exceptions for science and education, but these exceptions are much more limited than the fair use exceptions in copyright law and they are not required. For example, France does not permit any fair use exceptions in its database law. Other countries, including the United States, are currently considering similar database protections.

Contract

Most data agreements between users and data collectors (e.g., with the European Space Agency for use of ERS data[3]) are contracts. A contract is a two-party agreement, the terms of which are specified by the individuals involved. Because contracts can be written to circumvent fair

---

[2]For example, the Digital Millennium Copyright Act provides new rights and limitations related to the use of copyrighted works on the Internet or other digital environments. The act prohibits circumvention of technological measures to control access, even if it is done for legitimate reasons, and thus potentially reduces public access to information. See NRC, 2000, *The Digital Dilemma: Intellectual Property in the Information Age*. National Academy Press, Washington, D.C., 340 pp.

[3]Once a researcher's proposal to use ERS data is approved, the researcher signs a formal agreement naming every individual who would be permitted access to the data (see Example 5.5).

use, they are potentially a very powerful form of intellectual property protection. For example, contracts can be written to prevent multiple uses of the same data, sharing data with colleagues, or publishing the data on which the analysis was based. They can also be written to slow or prevent data from public-private partnerships from entering the public domain (e.g., Example 5.14). On the other hand, contract law has some limitations, including (1) a high administrative burden of negotiating terms with each user and provider of data, particularly for datasets compiled from several sources, and (2) they cannot fully prevent unauthorized downstream uses of the data because they are only binding on the parties to the agreement.

Appendix C

Acronyms

BAS	British Antarctic Survey
CNES	Centre National d'Etudes Spatiales
ENVISAT	ENVIronment SATellite
EROS	Earth Resources Observations Systems
ERS	European Remote-Sensing Satellite
ESA	European Space Agency
EU	European Union
IOC	Intergovernmental Oceanographic Commission
NASA	National Aeronautics and Space Administration
NEMO	Navy Earth Map Observer
NOAA	National Oceanic and Atmospheric Administration
NRC	National Research Council
OECD	Organization for Economic Cooperation and Development
OMB	Office of Management and Budget
SAR	Synthetic Aperture Radar
SeaWiFS	Sea-Viewing Wide Field-of-View Sensor
SPOT	Systeme Probatoire pour l'Observation de la Terre
STDC	Space Technology Development Corporation
TOMS	Total Ozone Mapping Spectrometer
TOPEX	Ocean TOPography EXperiment
TRMM	Tropical Rainfall Measuring Mission
UNESCO	United Nations Educational, Scientific, and Cultural Organization
USGCRP	U.S. Global Change Research Program
USGS	U.S. Geological Survey
WMO	World Meteorological Organization

GR 25 .N38 2001